Laboratory Mannual of
Plant Pathology

Vinod Kumar Jain

Oxford Book Company
Jaipur ▪ New Delhi

Published By

Oxford Book Company

Regd. Off. "Bony Residency", Gate No. 2
Opp. Tilak Public School, Vishweswariya Nagar,
Gopalpura Road, Jaipur - 302018, India
Ph: 0141 - 2761280, Telefax: 2761381
email - oxfordbookjpr@gmail.com
website - oxfordbookcompany.com

Delhi Office: 102, 1st Floor, Satyam House,
4327/3, Ansari Road, Darya Ganj,
New Delhi - 110002, India
Ph: 011-45652440

Distributed by
KEYA INTERNATIONAL PRESS LLP
E-mail: keyainternationalpressllp@gmail.com

Published in India, 2026

ISBN : 9788189473839

Typeset by Shivangi Computers, Jaipur

Printed and bound in India by Replika Press Pvt. Ltd.

Preface

As agriculture as an economic sector grows by leaps and bounds, an important factor has come into greater prominence than ever before: plant pathology. In a scenario where increased efficiency in both quality and quantity of the agricultural output has become necessary the concerns associated with plant diseases have also seen brought to the force. Plant pathology deals with the causes and processes of disease and its effects on the structure and function of plants. As such, plant pathologists are important since they specialise in interpreting the process of disease by examination of the tissues of the plant, and by determining the concentration of various biochemical sub stances within the plant body.

The present text has been intended as a handy and efficient laboratory manual for plant pathologists. Incorporating a variety of laboratory experiments in the field, it provides readers with the concepts, principles, techniques, applications and technology which is necessary to the practice of plant pathology in the laboratory for research and development. With the necessary illustrations and diagram also included, plant pathologists would be provided with a text not lacking in visual information crucial to pathology practice. In addition, the book manages to provide space for more recent experiments in the field.

Vinod Kumar Jain

Contents

1

Types of Plant Pathogens

Plant pathology (also called phytopathology) is the scientific study of plant diseases caused by pathogens (infectious diseases) and environmental conditions (physiological factors). Organisms that cause infectious disease include fungi, oomycetes, bacteria, viruses, viroids, virus-like organisms, phytoplasmas, protozoa, nematodes and parasitic plants. Not included are insects, mites, vertebrate or other pests that affect plant health by consumption of plant tissues.

The *"Disease triangle"* is a central concept of plant pathology for infectious diseases. It is based on the principle that disease is the result of an interaction between a host, a pathogen, and environment condition.

FUNGI

The majority of phytopathogenic fungi belong to the Ascomycetes and the Basidiomycetes. The fungi reproduce both sexually and asexually via the production of spores. These spores may be spread long distances by air or water, or they may be soil bourne. Many soil bourne spores, normally zoospores and capable of living saprotrophically, carrying out the first part of their lifecycle in the soil.

Fungal diseases can be controlled through the use of fungicides in agriculture, however new races of fungi often evolve that are resistant to various fungicides.

Fungal Plant Pathogens

Ascomycetes

Fusarium

Fusarium is a large genus of filamentous fungi widely distributed in soil and in association with plants. Most species are harmless saprobes and are relatively abundant members of the soil microbial community. Some species produce mycotoxins in cereal crops that can affect human and animal health if they enter the food chain. The main toxins produced by these *Fusarium* species are fumonisins and trichothecenes.

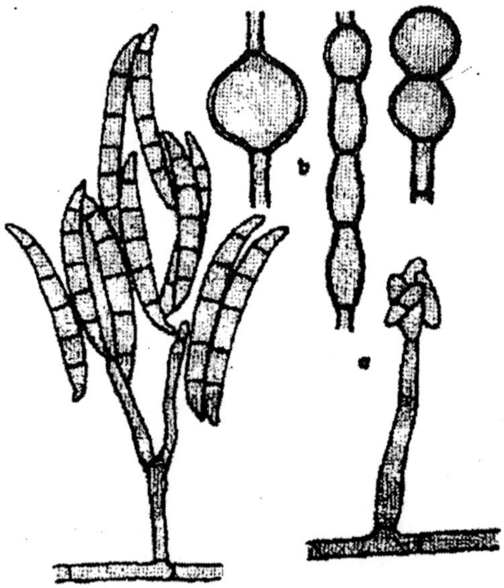

Fusarium

Fusarium graminearum commonly infects barley if there is rain late in the season. It is of economic impact to the malting and brewing industries as well as feed barley. Fusarium contamination in barley can result in head blight and in extreme contaminations the barley can appear pink.

The genome of this wheat and maize pathogen has been sequenced. *Fusarium graminearum* can also cause root rot and seedling blight. The total losses in the US of barley and wheat crops between 1991 and 1996 have been estimated at $3 billion.

Fusarium species is thought to cause the deadly white nose syndrome in bats, though it may be an opportunistic infection, rather than the actual cause of the condition.

Thielaviopsis

Thielaviopsis is a small genus of fungi in the order Microascales. The genus includes several important agricultural pathogens. The most widespread is *T. basicola,* the causal agent in several root rot diseases of economically important crop species including cotton and a variety of vegetables. In cotton, *Thielaviopsis* root rot, also known as black root rot causes necrosis of the roots and stunting of the crop plants.

Verticillium

Verticillium is a genus of fungi in the division Ascomycota. Within the genus, diverse groups are formed comprising saprotrophs and parasites of higher plants, insects, nematodes, mollusc eggs and other fungi thus it can be seen that the genus is a wide ranging group of taxa characterised by simple but ill-defined characters.

The genus may be broadly divided into three ecologically based groups 1) mycopathogens 2) entomopathogens and 3) plant pathogens and related saprotrophs. However, recently the genus has undergone some revision into which most entomopathogenic and mycopathogenic isolates fall into a new group called *Lecanicillium*. Plant pathogenic isolates still retain the original genus name *Verticillium*. The better known species of *Verticillium* are, *Vertcillium dahliae* and *Verticillium albo-*

atrum that cause wilt diseases in economically important plant species such as cotton, tomatoes, potatoes, eggplants, peppers, ornamental woody plants, as well as plants in natural vegetation communities.

Symptoms are superficially similar to Fusarium wilts. Crop rotation, the use of resistant crop varieties and deep plowing, may be useful in controlling *Verticillium* wilt.

The following are the selected species:

— *Verticillium dahliae* Kleb.

— *Lecanicillium lecanii* (approved name). Other names: Previously known as *Cephalosporium lecanii* Zimmermann and *Verticillium lecanii* (Zimmerman) Viegas. This fungus species was first described in 1861. It has a worldwide distribution found on insects. In horticulture and agriculture it is sometimes used as an entomopathogen (infects insects) for controlling insect pests like mealybugs and mites and aphids. It's a bio-pesticide/Bio-Insecticide that is sprayed on insects while they are feeding off of desirable plants. The insects are infected when they come into contact with the sticky fungal spores which then grow and invade the body, thus pathogenesized, the insects internal organs are consumed, leading to the insects death.

— *Verticillium albo-atrum Reinke & Berthold* - Causes Verticillium Wilt or Maple Wilt. First identified from potatoes in Germany in 1870. This species attacks over 300 different cultivated plants and can persist as a saprotrophic soil organism for more than 15 years. When infecting ornamental trees like Maples, Elms, Aspen, Ash, Beech, Catalpa, Oak plus many more the first symptoms are midsummer wilting on one side of a tree or branch. The sapwood has greenish or brownish streaks, and the infection can take a few years to progress to the rest of the tree or move rapidly. The fungi universally move up the xylem vessels. In fruit trees the infection is know as 'Black Heart' and is common in apricots and sometimes effects almonds, peaches, plums and avocado. This Fungus also effects herbaceous ornamentals and vegetables like Chrysanthemum, mints, Lychnis, tomatoes, eggplants, okra, rhubarb. Causing wilting and death. Identification can be made by looking for one-celled conidia, hyaline round to ellipsoid which are formed at the tips of whorled branches. They are easily separated from the tips.

Magnaporthe grisea

Magnaporthe grisea, also commonly known as rice blast fungus, rice rotten neck, rice seedling blight, blast of rice, oval leaf spot of graminea, pitting disease, ryegrass blast, and johnson spot, is a plant-pathogenic fungus that causes an important disease affecting rice. It can also infect a number of other agriculturally important cereals including wheat, rye, barley, and pearl millet causing diseases called blast disease or blight disease. *M. grisea* causes economically significant crop losses annually, each year it is estimated to destroy enough rice to feed more than 60 million people. The fungus is known to occur in 85 countries worldwide. It is sometimes thought of as a model organism in the study of phytopathogenic fungi.

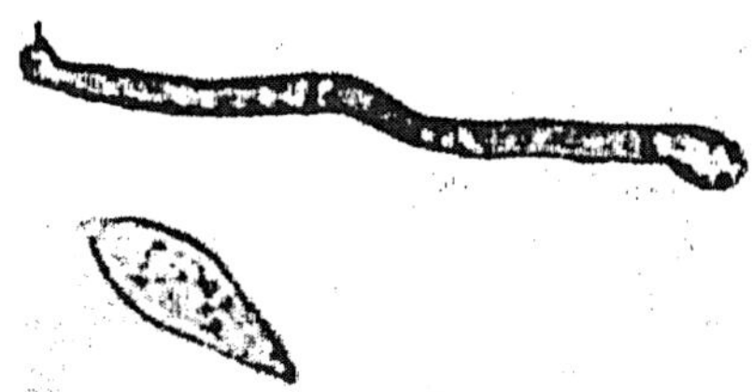

M. grisea is an ascomycete fungus. It is an extremely effective plant pathogen as it can reproduce both sexually and asexually to produce specialised infectious structures known as appressoria that infect aerial tissues and hyphae that can infect root tissues.

The asexual life cycle begins when the hyphae of the fungus undergo sporulation to produce fruiting structures called conidia which contain many spores. When these spores land on leaves and other aerial tissues of susceptible plants they germinate, developing the appressorium. The appressorium penetrates the plant cell by producing a penetration peg. Pressure in the appressorium increases and the structure explodes, forcing the penetration peg through the cell wall and into the cell. The fungus can then grow hyphae within the leaf and form lesions. Once established in the host plant the fungal hyphae can undergo asexual sporulation again. Sexual reproduction occurs when two strains of opposite mating types meet and form a perithecium in which ascospores develop. Once released, ascospores can develop appressoria and infect host cells. Spores are transmitted between plants by the wind.

In 2004 it was shown that in addition to infecting plants through the leaf, *M. grisea* can also infect the plant roots. The mode of root infection is the same as most root infecting fungi; it grows long hyphae that form an infection pad to gain entry to the root's interior. Once embedded in the root the fungus can produce resting structures. The blast fungus can also invade the plant's vascular system, growing inside the xylem and phloem and blocking the transport of nutrients and water from the roots, and produce lesions on aerial plant parts.

The fungus can kill or damage the plant in a number of ways. Aerial infection in seedlings is usually lethal because the young plants are unable to photosynthesize. Aerial infection in mature plants does not kill them, but reduced photosynthesis due to lesions on the leaf and use of photosynthate by the fungus greatly reduces yield. Infection of root and vascular tissues has the potential to kill the plant by cutting off the supply of water and nutrients to the root.

In 2005 the complete genome of *M. grisea* was sequenced. The organism is predicted to have over 11,000 genes. It is expected that the genome will reveal the mechanisms of fungal pathogen-plant interaction, in both aerial and root infection.

Rice blast was probably first recorded as *rice fever disease* in China in 1637. It was later described as *imochi-byo* in Japan in 1704, and as *brusone* in Italy in 1828. The fungus is currently reported to be present in at least 85 countries. In 1996 rice blast was found in rice in California, and has since been found in grasses on golf courses in the midwestern United States.

Strains of the fungus can infect domesticated grasses such as barley, wheat, rye, pearl millet, and turf grasses in addition to rice. Thus, even when crops are burned to destroy fungal infection, grass weeds can act as a disease reservoir. The disease may be called different names depending on the crop infected; In rice it is called rice blast, in wheat it is called wheat blast, in rye it is called rye blast and so on.

Control

M. grisea outbreaks are controlled through the application of expensive and potentially hazardous fungicides. Among the current blasticides are probenazole, tricyclazole, pyroquilon and phthalide. Infected crops are also burned in some areas.

With plant diseases, control is a relative term defining disease progress and subsequent loss of desirable plant product (rice grain here). The magnitude of control varies from very little to complete. With rice blast, efficacy of fungicides and plant genetic resistance is determined by the interaction of environmental and cultural conditions.

The fungus has been able to develop resistance to both chemical treatments and genetic resistance developed by plant breeders in some types of rice. It is thought that the fungus can achieve this by genetic change through mutation. Researchers hope that by having the full genome sequence of the fungus the development of effective control methods will be possible. For example, a mutant *M. grisea* unable to produce a penetration peg has been shown to lack the ability to infect rice.

Basidiomycetes

Rhizoctonia solani

Rhizoctonia solani (teleomorph: *Thanatephorus cucumeris*) is a plant pathogenic fungus with a wide host range and worldwide distribution. It is one cause of the condition known as damping off, which is a cause of death of seedlings in agriculture.

Rhizoctonia solani does not produce spores, hence it is identified only from mycelial characteristics. Its hyphal cells are multinucleate. Also, it produces white to deep brown mycelium when grown on artificial medium. The hyphae are 4-15 ìm wide and tend to branch at right angles. A septum near each hyphal branch and a slight constriction at the branch are diagnostic. *R. solani* is subdivided into anastomosis groups (AG) based on hyphal fusion between compatible strains.

The teleomorph of *R. solani* is *Thanatephorus cucumeris*. It forms club-shaped basidia with four apical sterigmata on which oval, hyaline basidiospores are borne.

Phakospora pachyrhizi

Soybean Rust (SRB) is caused by two related species of fungi, the most aggressive is Phakopsora pachyrhizi Sydow (anamorph Malupa sojae), and the less virulent species, Phakopsora meibomiae (Arthur) Arthur (anamorph Malupa vignae) which has only been found in limited areas in the Western Hemisphere and is not known to cause severe yield losses in soybean.

P. meibomiae, is referred to as the tropical, Latin American or New World rust strain. The Asian strain of soybean rust caused by *P. pachyrhizi* is extremely aggressive and is listed as a select biological agent, determined to have the potential to pose a threat to plant health or plant products.

Phakopsora pachyrhizi is an air-borne fungal pathogen that is not present in the continental United States. Under conducive environmental conditions, this pathogen could cause serious economic and crop losses in major soybean production regions of the United States.

Extreme weather events (such as a hurricane) carried *P. pachyrhizi* from infested areas to the United States. Long-distance dispersal of fungal spores by the wind can spread plant diseases across and even between continents and reestablish diseases in areas where host plants are seasonally absent. For such epidemics to occur, hosts that are susceptible to the same pathogen genotypes must be grown over wide areas, as is the case with many modern crops. Besides via wind current from Africa or South America, experts predict there are several ways spores could potentially be carried into this country. Those include: by tourists who have traveled internationally, via the movement of plant materials from infected countries, through the land bridge of Central America or even intentionally, through bioterrorism.

Urediospores germinate and the new uredinia matures in 10 days and they are capable of producing new urediospores for 3 weeks thereafter. The new urediospores may infect healthy plants, or an alternate or over-wintering host (such as kudzu). Urediniospores are the most common spore type found during the growing season. These urediniospores are the primary inoculum source and with prevailing winds and conducive environmental conditions can initiate soybean rust epidemic

P. pachyrhizi is still known as an obligate parasite. Uredospores survive (in resting or dormant state) less than 2 days under natural outdoor conditions. They do not survive in dried or decayed tissues or in the soil.

When soybean plants are not present in the field the fungus is viable only when transferred to another living host. One widespread host in the United States is kudzu, Pueraria montana var. lobata, that could serve as a reservoir for soybean rust. There are a variety of other important hosts that are leguminous crops or weeds that have shown varying degrees of susceptibility to both species of Soybean rust. Some other common hosts are yellow sweet clover (Melilotus officinalis), vetch (Vicia dasycarpa), medic (Medicago arborea), lupine (Lupinus hirsutus), green and kidney bean (Phaseolus vulgaris), lima and butter bean (Phaseolus lunatus), and cowpea or backeyed pea (Vigna unguiculata). Unfortunately, P. pachyrhizi infects and develops on kudzu but the weed is not adversely affected by the disease.

Infected hosts

Phakopsora pachyrhizi infects 31 species in 17 genera of legumes. Natural hosts (those hosts from which P. pachyrhizi has been found in situ include species in the Fabaceae family, sub-family Papilionoideae: Crotalaria spp., Desmodium spp., Glycine spp., Kennedia spp., Lablab purpureus, Lupinus spp., Macroptilium spp., Melilotus officinalis, Neonotonia wightii, Pachyrhizus erosus, Phaseolus spp., Pueraria lobata, Sesbania exaltata, Trifolium incarnatum, Vicia dasycarpa, Vigna spp. Sixty species in 26 other genera have been infected under controlled conditions. Any list of hosts requires cautious interpretation. In some cases a species may be designated "host" when the pathogen induces a visible symptom, regardless of whether the fungus sporulates in that species or not. From an epidemiological point of view, a species that does not support sporulation of the pathogen would be of no functional importance. Some plants permitting sporulation may be recorded as nonhosts because of an unexpectedly prolonged delay in the appearance of fertile uredinia.

The existence of pathotypes or races of the fungus and strains or varieties within a given legume species further complicates the picture. In other instances, results obtained in the field and those obtained in greenhouses and growth cabinet, leading to uncertainty about the appropriateness of designating a particular species or accession as a "host". From a practical point of view, the wild and cultivated legumes in each region where rust is a problem should be screened locally to endemic populations of P. pachyrhizi to determine their status as potential functional hosts. The information obtained will clarify the roles of these legumes as possible sources of initial inoculum, inoculum reservoirs, and overseasoning hosts. The unusually large host range may be due to the unique ability of the fungus to directly penetrate host epidermal cells through the formation of an appressorium and germ tube.

Puccinia

The stem, black or cereal rusts are caused by the fungus *Puccinia graminis* and are a significant disease affecting cereal crops. An epidemic of stem rust on wheat caused by race Ug99 is currently spreading across Africa, Asia and most recently into Middle East and is causing major concern.

There is considerable genetic diversity within the species *P. graminis* and several special forms, forma specialis, which vary in host range have been identified.

— Puccinia graminis f. sp. avenae, oat

— *Puccinia graminis f. sp. dactylis*

— *Puccinia graminis f. sp. lolii*

— *Puccinia graminis f. sp. poae*

— Puccinia graminis f. sp. secalis, rye, barley

— Puccinia graminis f. sp. tritici, wheat, barley

Like other Puccinia species, *P. graminis* has a complex life cycle featuring alternation of generations, the fungus is also heteroecious which means that its various life cycle stages require alternate host species. The complete life cycle of *P. graminis* requires barberry as well as a cereal species.

In the spring and summer, stem rust infections on wheat plants produce dikaryotic urediniospores, which are spread by the wind to nearby wheat plants, where they germinate and infect wheat by penetrating through the stomata. This polycyclic asexual phase can rapidly spread the infection over a wide area. Towards the end of the growing season, the rust converts to producing teliospores, which again contain these two haploid nuclei of opposite mating types. Before the winter, the nucleii fuse to form a diploid cell, which remains dormant until the next spring when it undergoes meiosis to produce four haploid cells known as basidiospores, borne on a structure called a basidium.

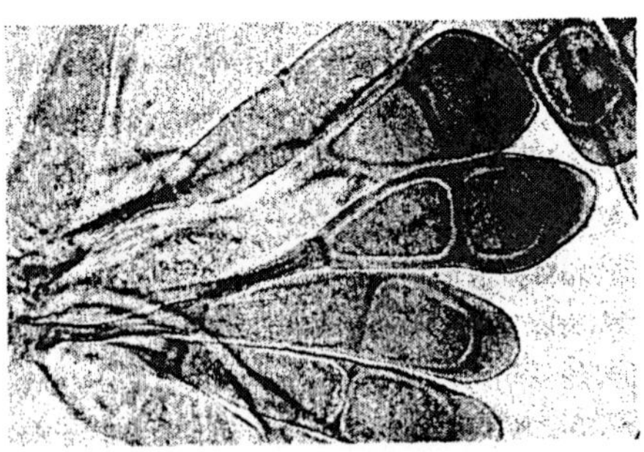

Puccinia graminis

The basidiospores then undergo a mitotic nuclear division to produce the mature basidiospore which contains two haploid nuclei of the same mating type. Basidiospores cannot infect cereal plants, but are instead, carried in the wind, and infect young leaves of common barberry (*Berberis vulgaris*) or other susceptible *Berberis*, Mahonia, or Mahoberberis species or cultivars. On barberry, the basidiospore penetrates the leaf epidermis directly, and the resulting infections produce specialised infection structures called pycnia (or spermagonia).

Pycnia (or spermagonia), which result from infection on young barberry leaves by basidiospores, are the sexual stage of the fungus life cycle. When a receptive hyphae from one pycnium has been fertilised by pycniospores (or spermatia) from a mating type compatible pycnium, its haploid cells become dikaryotic. The fertilised hyphae forms an aecium, on the underside of the barberry leaf, which produces chains of aeciospores surrounded by a bell-like enclosure of fungal cells. Like the urediniospores and like the cells of the aecium, each aeciospore contains two nuclei. Aeciospores are carried by the wind, and infect wheat by penetrating through stomata. After infecting a wheat plant, the aecispores develop and form uredia under the plants epidermis, these produce the dikaryotic urediniospores. These uredia eventually rupture the plants epidermis and again spread by the wind to nearby wheat plants, continuing the lifecycle.

Pathology

The stem rust fungus attacks the aboveground parts of the plant, infected plant produce fewer tillers and set fewer seed. In the case of extremely bad infection the plant may die. The site of infection is a visible symptom of the disease. Where infection has occurred on the stem or leaf, elliptical blisters or pustules called uredia develop

Pycnia typically form on the upper side of barberry leaves, and aecia form within 5-7 days after fertilisation on the lower side of the leaf directly below each fertilised pycnium.

Oomycetes

Oomycetes also known as Water molds are a group of filamentous, unicellular Heterokonts, physically resembling fungi. They are microscopic, absorptive organisms that reproduce both sexually and asexually and are composed of mycelia, or a tube-like vegetative body (all of an organism's mycelia are called its thallus). The name "water mold" refers to their earlier classification as fungi, which stemmed from their preference for conditions of high humidity and running surface water, although they are now known to have evolved separately and show a number of differences. For instance, their cell walls are composed of cellulose rather than chitin and generally do not have septations. Also, in the vegetative state they have diploid nuclei, whereas fungi have haploid nuclei.

Instead, water molds are related to organisms such as brown algae and diatoms, making up a group called the heterokonts. The name comes from the common arrangement and structure of motile cells, which typically have two unequal flagella. Among the water molds, these are produced as asexual spores called zoospores, which capitalise on surface water (including precipitation on plant surfaces) for movement. They also produce sexual spores, called oospores, that are translucent double-walled spherical structures used to survive adverse environmental conditions. A few produce aerial asexual spores that are distributed by wind.

The water molds are economically and scientifically important because they are aggressive plant pathogens. Some species can cause disease in fish. The majority can be broken down into three groups, although more exist.

— The *Phytophthora* group is a genus that causes diseases such as dieback, late blight in potatoes, sudden oak death, rhododendron root rot, and Ink Disease in the American Chestnut.

— The *Pythium* group is even more prevalent than Phytophythora and individual species have larger host ranges, usually causing less damage. Pythium damping off is a very common problem in greenhouses where the organism kills newly emerged seedlings. Mycoparasitic members of this group (e.g. *P. oligandrum*) parasitise other oomycetes and fungi, and have been employed as biocontrol agents. One Pythium species, Pythium insidiosum is also known to infect mammals.

— The third group are the downy mildews, which are easily identifiable by the appearance of white "mildew" on leaf surfaces (although this group can be confused with the unrelated powdery mildews).

Traditionally, this group was thought to include types of fungi, and indeed fungi themselves were once believed to be closely related to plants. However, further research has concluded that this is not true and that fungi are more closely related to animals. Many species of Oomycetes are still described or listed as types of fungi and may sometimes be referred to as pseudofungi, or lower fungi. Oomycetes are actually members of the chromistans, which are in turn part of the larger Kingdom Protoctista.

Significant oomycete plant pathogens are :

— *Pythium* spp.

— *Phytophthora* spp.; including the causal agent of the Great Irish Famine (1845-1849)

Pythium

Pythium is a genus of parasitic oomycete. Because this group of organisms were once classified as fungi, they are sometimes still treated as such.

Pythium, like others in the family *Pythiaceae,* are usually characterised by their production of coenocytic hyphae, hyphae without septations.

— *Oogonia :* Generally contain a single oospore
— *Antheridia:* Contain an elongated and club-shaped antheridium.

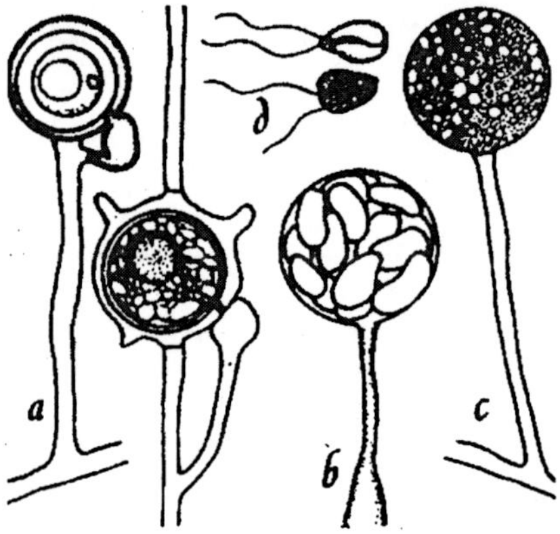

Pythium myriotylum

Pythium root rot is a common crop disease caused by a genus of organisms called "Pythium". These are commonly called water moulds. *Pythium* damping off is a very common problem in fields and greenhouses, where the organism kills newly emerged seedlings. This disease complex usually involves other pathogens such as Phytophthora and Rhizoctonia. Pythium wilt is caused by zoospore infection of older plants leading to biotrophic infections that become necrotrophic in response to colonization/reinfection pressures or environmental stress, leading to minor or severe wilting caused by impeded root functioning.

Many *Pythium* species, along with their close relatives, Phytophthora species are plant pathogens of economic importance in agriculture. *Pythium* spp. tend to be very generalistic and unspecific in their host range. They infect a large range of hosts, while *Phytophthora* spp. are generally more host-specific.

For this reason, *Pythium* spp. are more devastating in the root rot they cause in crops, because crop rotation alone will often not eradicate the pathogen (nor will fallowing the field, as *Pythium* spp. are also good saprotrophs, and will survive for a long time on decaying plant matter).

It has been noted that in field crops, damage by *Pythium* spp. is often limited to the area affected, as the motile zoospores require ample surface water to travel long distances. Additionally, the capillaries formed by soil particles act as a natural filter and effectively trap many zoospores. However, in hydroponic systems inside greenhouses, where extensive monocultures of plants are maintained in plant nutrient solution that is continuously recirculated to the crop, *Pythium* spp. cause extensive and devastating root rot and is often difficult to prevent or control. The root rot affects entire operations (tens of thousands of plants, in many instances) within two to four days due to the inherent nature of hydroponic systems where roots are nakedly exposed to the water medium, in which the zoospores can move freely.

Several *Pythium* species, including P. oligandrum, *P. nunn, P. periplocum,* and P. acanthicum are mycoparasites of plant pathogenic fungi and oomycetes, and have received interest as potential biocontrol agents.

Phytophthora

Phytophthora is a genus of plant-damaging Protists of the Oomycetes (water molds). Heinrich Anton de Bary described it for the first time in 1875.

Phytophthoras are mostly pathogens of dicotyledons, and are relatively host-specific parasites. Many species of

Phytophthora are plant pathogens of considerable economic importance. Phytophthora infestans was the infective agent of the potato blight that caused the Great Irish Famine. Plant diseases caused by this genus are difficult to control chemically, thus resistant cultivars are grown as a management strategy. Research beginning in the 1990s has placed some of the responsibility for European forest die-back on the activity of imported Asian *Phytophthoras*.

Other important *Phytophthora* diseases are:

- *Phytophthora alni* – causes alder root rot
- *Phytophthora cactorum* – causes rhododendron root rot affecting rhododendrons, azaleas and causes bleeding canker in hardwood trees
- *Phytophthora cinnamomi* - causes cinnamon root rot affecting woody ornamentals including arborvitae, azalea, Chamaecyparis, dogwood, forsythia, Fraser fir, hemlock, Japanese holly, juniper, Pieris, rhododendron, Taxus, white pine, and American chestnut
- *Phytophthora fragariae* - causes red root rot affecting strawberries
- *Phytophthora palmivora* - causes fruit rot in coconuts and betel nuts
- *Phytophthora ramorum* – infects over 60 plant genera and over 100 host species - causes Sudden Oak Death
- *Phytophthora quercina* – causes oak death
- *Phytophthora sojae* - causes soybean root rot

Phytophthora is sometimes referred to as a fungal-like organism but it is classified under a different kingdom altogether: Stramenopila. This is a good example of convergent evolution: *Phytophthora* is morphologically very similar to true Fungi yet its evolutionary history is quite distinct. In contrast to Fungi, stramenopiles are more closely related to plants than animals. Whereas Fungal cell walls are made primarily of chitin, stramenopile cell walls are constructed mostly of cellulose. Ploidy levels are different between these two kingdoms as are biochemical pathways.

Phytophthoras may reproduce sexually or asexually. In many species, sexual structures have never been observed, or have only been observed in laboratory matings. In homothallic species, sexual structures occur in single culture. Heterothallic species have mating strains, designated as A1 and A2.

When mated, antheridia introduce gametes into oogonia, either by the oogonium passing through the antheridium (amphigyny) or by the antheridium attaching to the proximal (lower) half of the oogonium (paragyny), and the union producing oospores. Like animals, but not like most true Fungi, meiosis is gametic, and somatic nuclei are diploid. Asexual (mitotic) spore types are chlamydospores, and sporangia which produce zoospores.

Chlamydospores are usually spherical and pigmented, and may have a thickened cell wall to aid in its role as a survival structure. Sporangia may be retained by the subtending hyphae (non-caducous) or be shed readily by wind or water tension (caducous) acting as dispersal structures. Also, sporangia may release zoospores, which have two unlike flagella which they use to swim towards a host plant.

BACTERIA

Most bacteria that are associated with plants are actually saprotrophic, and do no harm to the plant itself. However, a small number, around 100 species, are able to cause disease. Bacterial diseases are much more prevalent in sub-tropical and tropical regions of the world.

Most plant pathogenic bacteria are rod shaped (bacilli). In order to be able to colonise the plant they have specific pathogenicity factors. There are 4 main bacterial pathogenicity factors:

1. *Cell wall degrading enzymes* - used to break down the plant cell wall in order to release the nutrients inside. Used by pathogens such as Erwinia to cause soft rot.

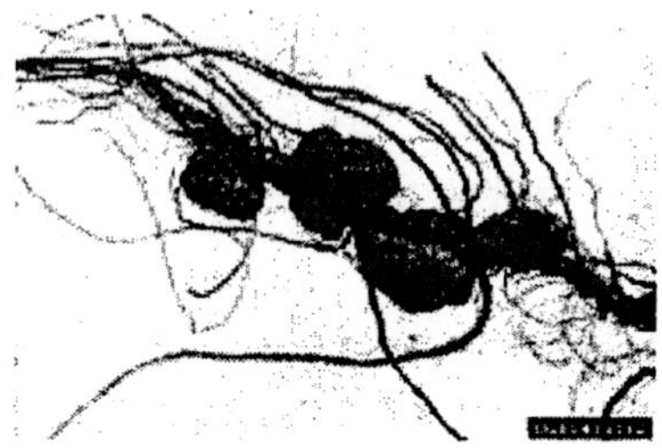

Crown gall disease caused by Agrobacterium

2. *Toxins* : These can be non-host specific, and damage all plants, or host specific and only cause damage on a host plant.
3. *Phytohormones* — for example Agrobacterium changes the level of Auxin to cause tumours.
4. *Exopolysaccharides*—these are produced by bacteria and block xylem vessels, often leading to the death of the plant.

Bacteria control the production of pathogenicity factors via quorum sensing.

Significant bacterial plant pathogens

— Proteobacteria
  — *Xanthomonas* spp.
  — *Pseudomonas* spp.

The Proteobacteria are a major group (phylum) of bacteria. They include a wide variety of pathogens, such as Escherichia, Salmonella, Vibrio, Helicobacter, and many other notable genera. Others are free-living, and include many of the bacteria responsible for nitrogen fixation. The group is defined primarily in terms of ribosomal RNA (rRNA) sequences, and is named for the Greek god Proteus (also the name of a bacterial genus within the Proteobacteria), who could change his shape, because of the great diversity of forms found in this group.

All Proteobacteria are Gram-negative, with an outer membrane mainly composed of lipopolysaccharides. Many move about using flagella, but some are non-motile or rely on bacterial gliding. The last include the myxobacteria, a unique group of bacteria that can aggregate to form multicellular fruiting bodies. There is also a wide variety in the types of metabolism. Most members are facultatively or obligately anaerobic and heterotrophic, but there are numerous exceptions. A variety of genera, which are not closely related to each other, convert energy from light through photosynthesis. These are called purple bacteria, referring to their mostly reddish pigmentation.

Cavalier-Smith has postulated that Proteobacteria is part of the bacterial clade Gracilicutes.

Alphaproteobacteria

The Alphaproteobacteria comprise most phototrophic genera, but also several genera metabolising C1-compounds (e.g. *Methylobacterium*), symbionts of plants (e.g. Rhizobia) and animals, and a group of dangerous pathogens, the Rickettsiaceae. Moreover the precursors of the mitochondria of eukaryotic cells are thought to have originated in this bacterial group.

Betaproteobacteria

The Betaproteobacteria consist of several groups of aerobic or facultative bacteria which are often highly versatile in their degradation capacities, but also contain chemolithotrophic genera (e.g. the ammonia-oxidising genus *Nitrosomonas*) and some phototrophs (genera *Rhodocyclus* and *Rubrivivax*). Beta Proteobacteria play an important role in nitrogen fixation in various types of plants, oxidizing ammonium to produce nitrite- an important chemical for plant function. Many of them are found in environmental samples, such as waste water or soil. Pathogenic species within this class are the Neisseriaceae (gonorrhea and meningoencephalitis) and species of the genus Burkholderia.

Gammaproteobacteria

The Gammaproteobacteria comprise several medically and scientifically important groups of bacteria, such as the Enterobacteriaceae, Vibrionaceae and Pseudomonadaceae. An exceeding number of important pathogens belongs to this class, e.g. Salmonella (enteritis and typhoid fever), Yersinia pestis (plague), Vibrio (cholera), Pseudomonas aeruginosa (lung infections in hospitalized or cystic fibrosis patients), and Escherichia coli O157:H7 (food poisoning).

Deltaproteobacteria

The Deltaproteobacteria comprise a branch of predominantly aerobic genera, the fruiting-body-forming myxobacteria, and a branch of strictly anaerobic genera, which contains most of the known sulfate- (*Desulfovibrio, Desulfobacter, Desulfococcus, Desulfonema*, etc.) and sulfur-reducing bacteria (e.g. *Desulfuromonas*) alongside several other anaerobic bacteria with different physiology (e.g. ferric iron-reducing Geobacter and syntrophic *Pelobacter* and *Syntrophus* species).

Epsilonproteobacteria

The Epsilonproteobacteria consist of only a few genera, mainly the curved to spirilloid Wolinella, Helicobacter, and Campylobacter. Most of the known species inhabit the digestive tract of animals and humans and serve as symbionts (*Wolinella* in cows) or pathogens (*Helicobacter* in the stomach, *Campylobacter* in the duodenum). There have also been numerous environmental sequences of epsilons recovered from hydrothermal vent and cold seep habitats.

Pseudomonas

Pseudomonas is a genus of gamma proteobacteria, belonging to the larger family of pseudomonads.

Recently, 16S rRNA sequence analysis has redefined the taxonomy of many bacterial species. As a result the genus *Pseudomonas* includes strains formerly classified in the genera *Chryseomonas* and *Flavimonas*. Other strains previously classified in the genus *Pseudomonas* are now classified in the genera Burkholderia and Ralstonia.

Pseudomonad literally means 'false unit', being derived from the Greek *pseudo* and *monas*. The term "monad" was used in the early history of microbiology to denote single-celled organisms.

Because of their widespread occurrence in water, the pseudomonads were observed early in the history of microbiology. The generic name *Pseudomonas* created for these organisms was defined in rather vague terms in 1894 as a genus of Gram-negative, rod-shaped and polar-flagella bacteria. Soon afterwards, a very large number of species was assigned to the genus. Pseudomonads were isolated from many natural niches and a large number of species names was originally assigned to the genus. New methodology and the inclusion of approaches based on the studies of conservative macromolecules have reclassified many strains.

Pseudomonas aeruginosa is increasingly recognized as an emerging opportunistic pathogen of clinical relevance. Several different epidemiological studies indicate that antibiotic resistance is increasing in clinical isolates.

In the year 2000, the complete genome sequence of a *Pseudomonas* species was determined; more recently the sequence of other strains have been determined including *P. aeruginosa* strains PAO1, *P. putida* KT2440, *P. fluorescens* Pf-5, *P. syringae* pathovar tomato DC3000, *P. syringae* pathovar syringae B728a, *P. syringae* pathovar phaseolica 1448A, *P. fluorescens* PfO-1 and *P. entomophila* L48.

Characteristics

Members of the genus display the following defining characteristics:

— Rod shaped
— Gram-negative
— One or more polar flagella, providing motility
— Aerobic, although some species have been found to be facultative anaerobes (e.g. *P. aeruginosa*)
— Non–spore forming
— Positive catalase test

Other characteristics which tend to be associated with *Pseudomonas* species include secretion of pyoverdin (fluorescein), a fluorescent yellow-green siderophore under iron-limiting conditions. Certain *Pseudomonas* species may also produce additional types of siderophore, such as pyocyanin by Pseudomonas aeruginosa and thioquinolobactin by Pseudomonas fluorescens. *Pseudomonas* species also typically give a positive result to the oxidase test, the absence of gas formation from glucose, glucose is oxidised in oxidation/fermentation test using Hugh and Leifson O/F test, beta hemolytic (on blood agar), indole negative, methyl red negative, Voges Proskauer test negative, citrate positive.

The genus demonstrates a great deal of metabolic diversity, and consequently are able to colonise a wide range of niches. Their ease of culture in vitro and availability of an increasing number of *Pseudomonas* strain genome sequences has made the genus an excellent focus for scientific research; the best studied species include *P.*

aeruginosa in its role as an opportunistic human pathogen, the plant pathogen *P. syringae,* the soil bacterium *P. putida,* and the plant growth promoting *P. fluorescens.*

Biofilm formation

All species and strains of *Pseudomonas* are Gram-negative rods, and have historically been classified as strict aerobes. Exceptions to this classification have recently been discovered in *Pseudomonas* biofilms. A significant number can produce exopolysaccharides that are known as slime layers. Secretion of exopolysaccharide makes it difficult for Pseudomonads to be phagocytosed by mammalian white blood cells. Slime production also contributes to surface-colonising biofilms which are difficult to remove from food preparation surfaces. Growth of Pseudomonads on spoiling foods can generate a "fruity" odor.

Pseudomonas have the ability to metabolise a variety of diverse nutrients. Combined with the ability to form biofilms, they are thus able to survive in a variety of unexpected places. For example, they have been found in areas where pharmaceuticals are prepared. A simple carbon source, such as soap residue or cap liner-adhesives is a suitable place for the Pseudomonads to thrive. Other unlikely places where they have been found include antiseptics such as quaternary ammonium compounds and bottled mineral water.

Antibiotic resistance

Being Gram-negative bacteria, most *Pseudomonas spp.* are naturally resistant to penicillin and the majority of related beta-lactam antibiotics, but a number are sensitive to piperacillin, imipenem, tobramycin, or ciprofloxacin.

This ability to thrive in harsh conditions is a result of their hardy cell wall that contains porins. Their resistance to most antibiotics is attributed to efflux pumps called ABC transporters, which pump out some antibiotics before they are able to act.

Pseudomonas aeruginosa is a highly relevant opportunistic pathogen. One of the most worrisome characteristics of *P. aeruginosa* consists in its low antibiotic susceptibility. This low susceptibility is attributable to a concerted action of multidrug efflux pumps with chromosomally-encoded antibiotic resistance genes and the low permeability of the bacterial cellular envelopes. Besides intrinsic resistance, *P. aeruginosa* easily develop acquired resistance either by mutation in chromosomally-encoded genes, either by the horizontal gene transfer of antibiotic resistance determinants. Development of multidrug resistance by *P. aeruginosa* isolates requires several different genetic events that include acquisition of different mutations and/or horizontal transfer of antibiotic resistance genes. Hypermutation favours the selection of mutation-driven antibiotic resistance in *P. aeruginosa* strains producing chronic infections, whereas the clustering of several different antibiotic resistance genes in integrons favours the concerted acquisition of antibiotic resistance determinants. Some recent studies have shown that phenotypic resistance associated to biofilm formation or to the emergence of small-colony-variants may be important in the response of *P. aeruginosa* populations to antibiotics treatment.

Taxonomy

The studies on the taxonomy of this complicated genus groped their way in the dark while following the classical procedures developed for the description and identification of the organisms involved in sanitary bacteriology during the first decades of the twentieth century. This situation sharply changed with the proposal to introduce as the central criterion the similarities in the composition and sequences of macromolecules components of the ribosomal

RNA. The new methodology clearly showed that the genus *Pseudomonas,* as classically defined, consisted in fact of a conglomerate of genera that could clearly be separated into five so-called rRNA homology groups. Moreover, the taxonomic studies suggested an approach that might proved useful in taxonomic studies of all other prokaryotic groups.

A few decades after the proposal of the new genus *Pseudomonas* by Migula in 1894, the accumulation of species names assigned to the genus reached alarming proportions. At the present moment, the number of species in the current list has contracted more than tenfold. In fact, this approximated reduction may be even more dramatic if one considers that the present list contains many new names, i.e., relatively few names of the original list survived in the process. The new methodology and the inclusion of approaches based on the studies of conservative macromolecules other than rRNA components, constitutes an effective prescription that helped to reduce *Pseudomonas* nomenclatural hypertrophy to a manageable size.

Pathogenicity

Animal pathogens

P. aeruginosa is an opportunistic human pathogen, most commonly affecting immunocompromised patients, such as those with cystic fibrosis or AIDS. Infection can affect many different parts of the body, but infections typically target the respiratory tract (e.g. patients with CF or those on mechanical ventilation), causing bacterial pneumonia. Treatment of such infections can be difficult due to multiple antibiotic resistance.

P. oryzihabitans can also be a human pathogen, although infections are rare. It can cause peritonitis, endophthalmitis, septicemia and bacteremia. Similar symptoms although also very rare can be seen by infections of *P. luteola.*

P. plecoglossicida is a fish pathogenic species, causing hemorrhagic ascites in the ayu (Plecoglossus altivelis). *P. anguilliseptica* is also a fish pathogen.

Due to their hemolytic activity, even non-pathogenic species of *Pseudomonas* can occasionally become a problem in clinical settings, where they have been known to infect blood transfusions.

Plant pathogens

P. syringae is a prolific plant pathogen. It exists as over 50 different pathovars, many of which demonstrate a high degree of host plant specificity. There are numerous other Pseudomonas species that can act as plant pathogens, notably all of the other members of the P. syringae subgroup, but P. syringae is the most widespread and best studied.

Although not strictly a plant pathogen, P. tolaasii can be a major agricultural problem, as it can cause bacterial blotch of cultivated mushrooms. Similarly, P. agarici can cause drippy gill in cultivated mushrooms.

Use as biocontrol agents

Since the mid 1980s, certain members of the *Pseudomonas* genus have been applied to cereal seeds or applied directly to soils as a way of preventing the growth or establishment of crop pathogens. This practice is generically referred to as biocontrol. The biocontrol properties of P. fluorescens strains are currently best understood, although it is not clear exactly how the plant growth promoting properties of *P. fluorescens* are achieved. Theories include: that the bacteria might induce systemic resistance in the host plant, so it can better resist attack by a true pathogen; the bacteria might out compete other (pathogenic) soil microbes, e.g. by siderophores giving a competitive advantage at scavenging for iron; the bacteria might produce compounds antagonistic to other soil microbes, such as phenazine-type antibiotics or hydrogen cyanide. There is experimental evidence to support all of these theories, in certain conditions; a good review of the topic is written by Haas and Defago.

Other notable *Pseudomonas* species with biocontrol properties include P. chlororaphis which produces a phenazine type antibiotic active agent against certain fungal plant pathogens, and the closely related species P. aurantiaca which produces di-2,4 diacetylfluoroglucylmethan, a compound antibiotically active against Gram-positive organisms.

Phytoplasmas ('Mycoplasma-like organisms') and spiroplasmas

Phytoplasma, formerly known as 'Mycoplasma-like organisms' or MLOs, are specialised bacteria that are obligate parasites of plant phloem tissue, and some insects. They were first discovered by scientists in 1967 when they were named mycoplasma like organisms or MLOs. They cannot be cultured in vitro in cell-free media. They are characterised by their lack of a cell wall, a pleiomorphic or filamentous shape, normally with a diameter less than 1 micrometer, and their very small genomes.

Phytoplasmas are pathogens of important crops, including coconuts and sugarcane, causing a wide variety of symptoms that ranges from mild yellowing to death of infected plants. They are most prevalent in tropical and sub-tropical regions of the world. Phytoplasmas require a vector to be transmitted from plant to plant and this normally takes the form of sap sucking insects such as leaf hoppers in which they are also able to replicate.

There are references to diseases known to now be caused by phytoplasmas as far back as 1603 for Mulberry dwarf disease in Japan. Such diseases were originally thought to be caused by viruses which are require insect vectors, can not be cultured and have some symptom similarity. In 1967 phytoplasmas were discovered in ultrathin sections of plant phloem tissue and named mycoplasma-like organisms due to the fact they physically resembled mycoplasmas The organisms were renamed phytoplasmas in 1994 at the 10th congress of the International Organisation of Mycoplasmology.

Morphology

Being mollicutes, phytoplasmas lack cell walls and instead are bound by a triple layered membrane. The cell membranes of all phytoplasmas studied so far usually contain a single immunodominant protein (of unknown function) that makes up the majority of the protein content of the cell membrane. Their shape is normally pleiomorphic or filamentous and normally have a diameter of less than 1 micrometer. Like other prokaryotes, DNA is free in the cytoplasm. They are believed to reproduce through Binary fission.

Symptoms

A common symptom caused by phytoplasma infection is phyllody, the production of leaf like structures in place of flowers. Evidence suggests that the phytoplasma downregulates a gene involved in petal formation (*AP3* and its orthologues) and genes involved in the maintenance of the apical meristem (*Wus* and *CLV1*). This causes sepals to form where petals should. Other symptoms, such as the yellowing of leaves, are thought to be caused by the phytoplasma's presence in the phloem affecting its function, and changing the transport of carbohydrates.

Phytoplasma infected plants may also suffer from virescence - the development of green flowers due to the loss of pigment in the petal cells. Sometimes sterility of the flowers is also seen.

Many phytoplasma infected plants gain a bushy or witch's broom appearance due to changes in normal growth patterns caused by the infection. Most plants show apical dominance but phytoplasma infection can cause the proliferation of auxiliary (side) shoots and an increase in size of the internodes. Such symptoms are actually useful in the commercial production of poinsettia. The infection is necessary to produce more axillary shoots that enable to production of pionsettia plants that have more than one flower.

Phytoplasmas may cause many other symptoms that are induced because of the stress placed on the plant by infection rather than specific pathogenicity of the phytoplasma. Photosynthesis, especially photosystem II, is inhibited in many phytoplasma infected plants. Phytoplasma infected plants often show yellowing which is caused by the breakdown of chlorophyll, whose biosynthesis is also inhibited.

Movement between plants

The phytoplasmas are mainly spread by insects of the families Cicadellidea (leafhoppers) and Fulgoridea (planthoppers) which feed on the phloem tissues of infected plants picking up the phytoplasmas and transmitting them to the next plant they feed on. For this reason the host range of phytoplasmas is strongly dependent upon its insect vector. Phytoplasmas contain a major antigenic protein that makes up the majority of their cell surface proteins and this has been shown to interact with insect microfilament complexes and is believed to the determining factor is insect-phytoplasma interation. Phytoplasmas may overwinter in insect vectors or perennial plants. Phytoplasmas can have varying effects on their insect hosts, examples of both reduced and increased fitness have been seen.

Phytoplasmas will be found in most of the major organs of an infected insect host once they are established. They will enter the insect's body through the stylet and then move through the intestine and be absorbed into the haemolymph. From here they proceed to colonise the salivary glands, a process that can take up to three weeks. The time between phytoplasmas being taken up by the insect and the phytoplasmas reaching an infectious titre in the salivary gland is called the latency period.

Phytoplasmas can also be spread via vegetative propagation such as the grafting of a piece of infected plant onto a healthy plant.

Movement within plants

Phytoplasmas are able to move within the pholem from source to sink and they are able to pass through sieve tube elements, but spread more slowly than solutes, for this and other reasons movement by passive translocation is not supported.

Detection and diagnosis

Before molecular techniques were developed the diagnosis of phytoplasma diseases was difficult because they could not be cultured. Thus classical diagnostic techniques such as observation of symptoms were used. Ultrathin sections of the phloem tissue from suspected phytoplasma infected plants would also be examined for their presence. Another diagnostic technique used was to treat infected plants with antibiotics such as tetracycline to see if this cured the plant.

Molecular diagnostic techniques for the detection of phytoplasma began to emerge in the 1980s and included ELISA based methods. In the early 1990s PCR-based methods were developed that were far more sensitive than those that used ELISA, and RFLP analysis allowed the accurate identification of different strains and species of phytoplasma.

There are also techniques that allow the assessement of the level of infection. Both QPCR and bioimaging have been shown to be effective methods of quantifying the titre of phytoplasmas within the plant.

Control

Phytoplasmas are normally controlled by the breeding and planting of disease resistance varieties of crops (believed to the most economically viable option) and by the control of the insect vector.

Tissue culture can be used to produce clones of phytoplasma infected plants that are healthy. The chances of gaining healthy plants in this manner can be enhanced by the use of cryotherapy, freezing the plant samples in liquid nitrogen before using them for tissue culture.

Work has also been carried out investigating the effectiveness of plantibodies targeted against phytoplasmas.

Tetracyclines are bacteriostatic to phytoplasmas, that is they inhibit their growth. However without continuous use of the antibiotic disease symptoms will reappear. Thus tetracycline is not a viable control agent in agriculture, but is used to protect ornamental coconut trees.

Viruses, Viroids and Virus-like Organisms

Plant viruses are viruses affecting plants. Plant viruses, like all other viruses, are obligate intracellular parasites that do not have the molecular machinery to replicate without the host. The plant viruses are defined as viruses pathogenic to higher plants. While this article does not intend to list all plant viruses, it discusses some important viruses as well as their uses in plant molecular biology.

Although plant viruses are not nearly as well understood as the animal counterparts, one plant virus has become iconic. The first virus to be discovered was *Tobacco mosaic virus* (TMV). This and other viruses cause have an estimated US$60 billion per year economic influence on crops worldwide. Plant viruses are grouped into 73 genera and 49 families.

The discovery of plant viruses causing disease is often accredited to Martinus Beijerinck who determined, in 1898, that plant sap obtained from tobacco leaves with the "mosaic disease" remained infectious when passed through a porcelain filter. This was in contrast to bacteria microorganisms, which were retained by the filter. Beijerinck referred to the infectious filtrate as a "contagium vivum fluidum", thus the coinage of the modern term "virus".

After the initial discovery of the 'viral concept' there was need to classify any other known viral diseases based on the mode of transmission even though microscopic observation proved fruitless. In 1939 Holmes published a classification list of 129 plant viruses. This was expanded and in 1999 there were 977 officially recognised, and some provisional, plant virus species.

The purification (crystallisation) of the TMV was first performed by Wendell Stanley, who published his findings in 1935, although he did not determine that the RNA was the infectious material. However, he received the Nobel Prize in Chemistry in 1946. In the 1950s a discovery by two labs simultaneously proved that the purified RNA of the TMV was infectious which reinforced the argument. The RNA was carrying genetic information to code for the production of new infectious particles.

More recently virus research has been focused on understanding the genetics and molecular biology of plant virus genomes, with a particular interest in determining how the virus can replicate, move, and infect plants. Understanding the virus genetics and protein functions has been used to explore the potential for commercial use by biotechnology companies. In particular, viral-derived sequences have been used to provide an understanding of novel forms of resistance. The recent boom in technology allowing humans to manipulate plant viruses may provide new strategies for production of value-added proteins in plants.

Structure

Viruses are very small and can only be observed with an electron microscope. The structure of a virus is given by its coat of proteins, which surround the viral genome. Assembly of viral particles takes place spontaneously.

Over 50% of known plant viruses are rod shaped (flexuous or rigid). The length of the particle is normally dependent on the genome but it is usually between 300–500 nm with a diameter of 15–20 nm. Protein subunits can be placed around the circumference of a circle to form a disc. In the presence of the viral genome, the discs are stacked, then a tube is created with room for the nucleic acid genome in the middle.

The second most common structure amongst plant viruses are isometric particles. They are 40–50 nm in diameter. In cases when there is only a single coat protein, the basic structure consists of 60 T subunits, where T is an integer. Some viruses may have 2 coat proteins are the formation of the particle is analogous to a football.

There are three genera of Geminiviridae that possess geminate particles which are like two isometric particles stuck together.

A very small number of plant viruses have, in addition to their coat proteins, a lipid envelope. This is derived from the plant cell membrane as the virus particle buds off from the cell.

Transmission of plant viruses

Through sap

It implies direct transfer of sap by contact of and wounded

plant with a healthy one. Such process occurs during agricultural practices by tools, hands, or by animal feeding on the plant. Generally TMV, potato viruses and cucumber mosaic viruses are transmitted via sap.

Insects

Plant viruses need to be transmitted by a vector, most often insects such as leafhoppers. One class of viruses, the Rhabdoviridae, have been proposed to actually be insect viruses that have evolved to replicate in plants. The chosen insect vector of a plant virus will often be the determining factor in that virus' host range: it can only infect plants that the insect vector feeds upon. This was shown in part when the old world white fly made it to the USA, where it transferred many plant viruses onto new hosts.

Depending on the way they are transmitted, plant viruses are classified as non-persistent, semi-persistent and persistent. In non-persistent transmission, viruses become attached to the distal tip of the stylet of the insect and on the next plant it feeds on, it inoculates it with the virus. Semi-persistent viral transmission involves the virus entering the foregut of the insect.

Those viruses that manage to pass through the gut into the haemolymph and then to the salivary glands are known as persistent. There are two sub-classes of persistent viruses: propergative and circulative. Propergative viruses are able to replicate in both the plant and the insect (and may have originally been insect viruses), whereas circulative can not.

Many plant viruses encode within their genome polypeptides with domains essential for transmission by insects. In non-persistent and semi-persistent viruses, these domains are in the coat protein and another protein known as the helper component. A bridging hypothesis has been proposed to explain how these proteins aid in insect-mediated viral transmission. The helper component will bind to the specific domain of the coat protein, and then the insect mouthparts — creating a bridge.

In persistent propagative viruses, such as tomato spotted wilt virus (TSWV), there is often a lipid coat surrounding the proteins that is not seen in the other classes of plant viruses. In the case of TSWV, 2 viral proteins are expressed in this lipid envelope. It has been proposed that the viruses bind via these proteins and are then taken into the insect cell by receptor-mediated endocytosis.

Nematodes

Soil-borne nematodes also have been shown to transmit viruses. They acquire and transmit them by feeding on infected roots. Viruses can be transmitted by non-persistently and persistently, but there is no evidence of viruses being able to replicate in nematodes.

Plasmodiophorids

A number of viral genera are transmitted, both persistently and non-persistently, by soil borne zoosporic protozoa. These protozoa are not phytopathogenic themselves, but parasitic. Transmission of the virus takes place when they become associated with the plant roots.

An example is Polymyxa graminis, which has been shown to transmit a plant viral diseases in ceral crops.

Seed and pollen borne viruses

Plant virus transmission from generation to generation occurs in about 20% of plant viruses. When viruses are transmitted by seeds, the seed is infected in the generative cells and the virus is maintained in the germ cells and sometimes, but less often, in the seed coat. When the growth and development of plants is delayed because of situations like unfavourable weather, there is an increase in the amount of virus infections in seeds. There does not seem to be a correlation between the location of the seed on the plant and its chances of being infected. Little is known about the mechanisms involved in the transmission of plant viruses via seeds, although it is known that it is environmentally influenced and that seed transmission occurs because of a direct invasion of the embryo via the ovule or by an indirect route with an attack on the embryo mediated by infected gametes. These processes can occur concurrently or separately depending on the host plant. It is unknown how the virus is able to directly invade and cross the embryo and boundary between the parental and progeny generations in the ovule. Many plants species can be infected through seeds including but not limited to the families Leguminoseae, Solanacease, Compositae, Rosaceae, Curcurbitaceae, Gramineae.

Translation of Plant Viral Proteins

As mentioned above, 90% of plant viruses have genomes that consist of single stranded RNA, meaning that they are in the same sense orientation as messenger RNA. Viruses use the plant ribosomes to produce the 4-10 proteins encoded by their genome. However, since all of the proteins are encoded on a single strand (that is, they are ploycistronic) this will mean that the ribosome will either only produce one protein, as it will terminate translation at the first stop codon or that a polyprotein will be produced. Plant viruses have had to evolve special techniques to allow the production of viral proteins by plant cells.

5' Cap

In order for translation to occur eukaryotic mRNAs require a 5' Cap structure. This means that viruses must also have one. This normally consists of 7MeGpppN where N is normally adenine or guanine. The viruses encode a protein, normally a replicase, with a methyltransferase activity to allow this.

Some viruses are cap-snatchers. During this process, a ^{7m}G-capped host mRNA is recruited by the viral transcriptase complex and subsequently cleaved by a virally encoded endonuclease. The resulting capped leader RNA is used to prime transcription on the viral genome.

However some plant viruses do not use cap, yet translate efficiently due to cap-independent translation enhancers present in 5' and 3' untranslated regions of viral mRNA.

Production of sub-genomic RNAs

Some viruses use the production of sub-genomic RNAs to ensure the translation of all proteins within their genomes. In this process the first protein encoded on the genome, and this the first to be translated, is a replicase. This protein will act of the rest of the genome producing negative strand sub-genomic RNAs then act upon these to form positive strand sub-genomic RNAs that are essentially mRNAs ready for translation.

Segmented genomes

Some viral families, such as the Bromoviridae instead opt to have multi-partite genomes, genomes split between multiple viral particles. For infection to occur, the plant must be infected with all particles across the genome. For instance Brome mosaic virus has a genome split between 3 viral particles, and all 3 particles with the different RNAs are required for infection to take place.

Polyprotein processing

This stratergy is adopted by viral genra such as the Potyviridae and Tymovirus. The ribosome translates a single protein from the viral genome. Within the polyprotein is an ezyme with proteinase function that is able cleave the polyprotein into the various single proteins or just cleave away the replicase, which can then produce sub-genomic RNAs.

Nematodes

Nematodes are small, multicelluar wormlike creatures. Many live freely in the soil, but there are some species which parasitise plant roots. They are mostly a problem in tropical and subtropical regions of the world, where they may infect crops. Root knot nematodes have quite a large host range, whereas cyst nematodes tend to only be able to infect a few species. Nematodes are able to cause radical changes in root cells in order to facilitate their lifestyle.

Protozoa

There are a few examples of plant diseases caused by protozoa. They are transmitted as zoospores which are very durable, and may be able to survive in a resting state in the soil for many years. They have also been shown to transmit plant viruses.

When the motile zoospores come into contact with a root hair they produce a plasmodium and invade the roots.

Parasitic plants

Parasitic plants such as mistletoe and dodder are included in the study of phytopathology. Dodder, for example, is used as a conduit for the transmission of virues or virus-like agents from a host plant to either a plant that is not typically a host or for an agent that is not graft-transmissible.

REFERENCES

George N. Agrios, *Plant Pathology*, fourth edition, Academic Press, New York, 1997.

Purcell, A. H., "Patterns and processes of plant diseases and insect vectors", *Agriculture and Forestry Bulletin of the University of Alberta 5(3):19-22.*

2

Plant Disease Epidemiology

Plant Disease epidemiology is the study of disease in plant populations. Much like diseases of humans and animals, plant diseases occur due to pathogens such as bacteria, viruses, fungi, oomycetes, nematodes, phytoplasmas, protozoa, and parasitic plants. Plant disease epidemiologists strive for an understanding of the cause and effects of disease and develop strategies to intervene in situations where crop losses may occur. Typically successful intervention will lead to a low enough level of disease to be acceptable, depending upon the value of the crop.

Plant disease epidemiology is often looked at from a multi-disciplinary approach, requiring biological, statistical, agronomic and ecological perspectives. Biology is necessary for understanding the pathogen and its life cycle. It is also necessary for understanding the physiology of the crop and how the pathogen is adversely affecting it. Agronomic practices often influence disease incidence for better or for worse.

Ecological influences are numerous. Native species of plants may serve as reservoirs for pathogens that cause disease in crops. Statistical models are often applied in order to summarise and describe the complexity of plant disease epidemiology, so that disease processes can be more readily understood. For example, comparisons between patterns of disease progress for different diseases, cultivars, management strategies, or environmental settings can help in determining how plant diseases may best be managed. Policy can be influential in the occurrence of diseases, through actions such as restrictions on imports from sources where a disease occurs.

ELEMENTS OF EPIDEMIC

Disease epidemics in plants can cause huge losses in yield of crops as well threatening to wipe out an entire species such as was the case with Dutch Elm Disease and could occur with Sudden Oak Death. An epidemic of potato late blight, caused by Phytophthora infestans, led to the Great Irish Famine and the loss of many lives.

Commonly the elements of an epidemic are referred to as the "disease triangle": a susceptible host, pathogen, and conducive environment. For disease to occur all three of these must be present. Below is an illustration of this point. Where all three items meet there is disease. The fourth element missing from this illustration for an epidemic to occur, is time. As long as all three of these elements are present disease can initiate, an epidemic will only ensue if all three continue to be present. Any one of the three might be removed from the equation though. The host might outgrow susceptibility as with high-temperature adult-plant resistance, the environment changes and is not conducive for the pathogen to cause disease, or the pathogen is controlled through a fungicide application for instance.

Sometimes a fourth factor of time is added as the time at which a particular infection occurs, and the length of time conditions remain viable for that infection, can also play an important role in epidemics. The age of the plant species can also play a role, as certain species change in their levels of disease resistance as they mature; a process known as ontogenic resistance.

If all of the criteria are not met, such as a susceptible host and pathogen are present but the environment is not conducive to the pathogen infecting and causing disease, disease cannot occur. For example, corn is planted into a field with corn residue that has the fungus *Cercospora zea-maydis*, the causal agent of Grey leaf spot of corn, but if the weather is too dry and there is no leaf wetness the spores of the fungus in the residue cannot germinate and initiate infection.

Likewise, it stands to reason if the host is susceptible and the environment favours the development of disease but the pathogen is not present there is no disease. Taking the example above, the corn is planted into a ploughed field where there is no corn residue with the fungus *Cercospora zea-maydis*, the causal agent of Grey leaf spot of corn, present but the weather means long periods of leaf wetness, there is no infection initiated.

When a pathogen requires a vector to be spread then for an epidemic to occur the vector must be plentiful and active. "Epidemic" also is commonly used to denote a sudden and rapid or a widespread development of disease, implying that if development is slow or it is spatially limited there is no epidemic.

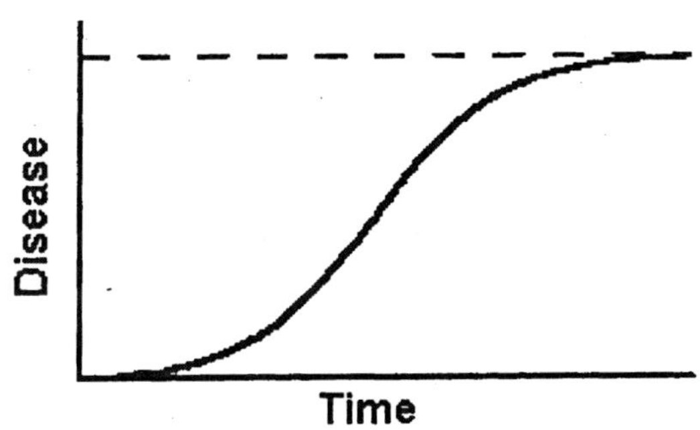

Dynamics of Plant Disease

Epidemic denote the dynamics of disease, that is, the change in the amount of disease with time. By "amount" of disease we mean either incidence (the number of plants or plant parts affected), severity (the amount of tissue affected), or a combination of both. By this definition there can be fast epidemics or slow epidemics, and there can even be "negative" epidemics, in which there is a decrease in the amount of disease with time. For example, if we are using the proportion of leaves infected as a measure of disease incidence, and during a dry spell we get new, healthy leaves forming while the infected ones drop off, we could get a negative change in the amount of disease.

Associated with the change in the amount of plant disease with time is a change in the spatial distribution of lesions or of infected plants. A good working definition of epidemic, therefore, is a change in plant disease in time and space.

The term endemic is sometimes used (erroneously) to mean the opposite of "epidemic", that is, to denote constant disease, because in areas where the pathogen and its host have coevolved over long periods of time, they often have reached an equilibrium where there is little change in the amount of disease with time. In the strict sense "endemic" means native to a particular area, and therefore we can have epidemics of endemic diseases.

TYPES OF EPIDEMICS

Monocyclic epidemics are caused by pathogens with a low birth rate and death rate meaning they only have one infection cycle per season. They are typical of soil born diseases such as Fusarium wilt of flax. Polycyclic epidemics are caused by pathogens capable of several infection cycles a season. These are most often caused by airborne diseases such as powdery mildew. Bimodal polycyclic epidemics can also occur. For example in brown rot of stone fruits the blossoms and the fruits are infected at different times.

For some diseases it is important to consider the disease occurrence over several growing seasons, especially if growing the crops in monoculture year after year or growing perennial plants. Such conditions can mean that the inoculum produced in one season can be carried over to the next leading to a build of an inoculum over the years. In the tropics there are no clear cut breaks between growing seasons as there are in temperate regions and this can lead to accumulation of innoculum.

Epidemics that occur under these conditions are referred to as polyetic epidemics and can be caused by both monocylcic and polycyclic pathogens. Apple powdery mildew is an example of a polyetic epidemic caused by a polycyclic pathogen and Dutch Elm disease a polyetic epidemic caused by a monocyclic pathogen.

DEVELOPMENT OF DISEASE

Disease on plants usually starts out at a low level, a small number of plants affected and a small amount of plant tissue affected, and it becomes of concern to us only when its incidence and severity increases with time. For example, in Phytophthora blight of pepper seedlings (*Phytophthora capsici*) and Fusarium kernel rot (*Fusarium moniliforme, currently F. verticilloides*) of maize, disease progress is roughly linear (allowing for some minor deviations that we can consider random error).

On the other hand, in bean rust (*Uromyces phaseoli*) and grey leaf spot of corn (*Cercospora zeae-maydis*), there is a definite upward curve; that is, disease increases at an increasing rate, a curve we could call exponential.

Obviously plant disease cannot continue to increase forever, and as the level of disease approaches 100%, the disease progress curve gradually flattens out. For example, in epidemics such as the infection of beans by Sclerotium rolfsii or the infection of tobacco by *Phytophthora nicotianae,* disease progress starts out looking linear but slows down as it approaches a maximum.

Likewise, the disease progress curves of *Puccinia graminis* subsp. *graminicola* on ryegrass and *Pyrenophora teres* f. sp. *teres* on barley appear exponential at first, but as time goes on and the incidence and severity of disease approach 100%, the rate of disease progress gradually slows to zero, giving both curves a somewhat sigmoid shape.

To be sure, not all examples of disease progress can be as neatly categorised as these, but in general plant disease epidemics tend to be either roughly linear or exponential in the early stages, and they tend to level off as they approach some limit.

The impact of plant disease and the losses that it causes are a function of disease progress. To reduce this impact, we need not eliminate the disease, we merely need to keep disease development below an acceptable level. That means that the progress of disease and the factors that influence disease progress must be understood in quantitative terms. We have to know what kinds of diseases lead to linear disease progress and what factors affect the slope of the line (the rate of disease progress). Likewise, we have to know what kinds of diseases tend to produce exponential disease progress curves and how we can reduce both the starting level of disease and the rate of epidemic development. Finally, we have to know why epidemics sometimes level off and what imposes limits to their development.

CYCLICAL NATURE OF PLANT DISEASE

Plant disease epidemics are cyclical phenomena, that is, they consist of repeated cycles of pathogen development in relation to the host and the environment. The inoculum, which might consist of fungal spores, bacterial cells, nematodes, viruses within an aphid vector, or some other propagules of a pathogen, gains entry into and establishment within the host tissues through the process of infection. The pathogen develops within the host and eventually begins to produce new inoculum, which, in turn, can be dispersed to new susceptible sites to initiate new infections. Pathogens that produce only one cycle of development (one infection cycle) per crop cycle are called monocyclic, while pathogens that produce more than one infection cycle per crop cycle are called polycyclic.

Generally in temperate climates there is only one crop cycle per year, so the terms "monocyclic" and "polycyclic" are based on the number of cycles per year. In tropical or subtropical climates, however, there can be more than one crop cycle per year, and it is important to remember that "monocyclic" and "polycyclic" are based on a single crop cycle. These same terms are used to describe the epidemics as well as the pathogens, so we often speak of a "monocyclic epidemic" or a "polycyclic epidemic".

For some diseases it is important to consider an epidemic over a period of many growing seasons. This is particularly true for perennial plants (forages, pastures, lawns, orchards, forests, etc.) or for annual crops that are grown in monoculture year after year. In these situations the inoculum produced in one growing season carries over to the next, and there could actually be a buildup of inoculum over a period of years. In the tropics there may not be clear-cut breaks between growing seasons such as we find in the temperate zones, and epidemics can be virtually continuous over periods of many years on such crops as bananas, coffee, and rubber trees. We refer to these epidemics as polyetic epidemics, regardless of whether the pathogen is monocyclic or polycyclic within each season.

The Dutch elm disease is an example of a monocyclic pathogen giving rise to a polyetic epidemic. Note that while there is only one cycle of infection each year and disease progress within each year is roughly linear, the incidence of infected trees increases at an increasing rate from year to year.

Apple powdery mildew is an example of a polyetic epidemic caused by a polycyclic pathogen. Note how the incidence of infection at the start of each year tends to increase exponentially.

Sigatoka leaf spot on bananas decreases somewhat during the dry season but otherwise produces more or less continuously repeated cycles of infection. The host, in this case, consists of a population of plants of different ages developing continuously over a long period of time.

Disease Cycles

Monocyclic Epidemics

In general, there are three types of plant diseases that tend to produce only one infection cycle per host cycle (1) postharvest diseases, (2) diseases caused by soil-borne plant pathogens, and (3) rusts without a urediniospore stage.

Postharvest diseases

Not all postharvest diseases produce monocyclic epidemics, but in many cases the infections that result in storage rots have either already occurred before harvest or occur during the harvest and postharvest handling before the product goes into storage. The rot progresses during storage, and new inoculum may be continuously produced, but unless the stored product is handled again to disperse the inoculum and create new infection sites by making small wounds, there are no new infection cycles. Note that the pathogen is not inherently monocyclic but is constrained by the environment to produce only a single cycle of infection. In other environments, these same pathogens might produce polycyclic epidemics.

Diseases caused by soilborne plant pathogens

Many of the root rots, vascular wilts, and other diseases caused by soilborne pathogens also produce only one infection cycle per crop cycle. The inoculum generally is some kind of survival structure resistant to desiccation or freezing, such as sclerotia, chlamydospores, or oospores in the soil or mycelium in crop residues. This inoculum is dispersed in the soil by plowing and fitting the land and turning under the crop residues. As the roots of the newly planted crop grow through the soil, they encounter the propagules of the pathogen embedded in the soil matrix and become infected. The epidemic progresses as new infections occur, but since any new inoculum that is produced is not dispersed until the soil is again cultivated, there is only one complete infection cycle per cropping cycle. To be sure, not all soilborne plant pathogens produce monocyclic epidemics, and one must be very careful to understand the life cycle of each pathogen before drawing conclusions about its epidemiology.

Demicyclic rusts

Some rusts produce no urediniospore stage (repeating stage) on a single host, and the inoculum produced on one host species generally must infect a different host species. This alternation of hosts appears to have evolved in adaptation to the annual growth cycles of the hosts, and we see one infection cycle on each host per year. An example is the cedar-apple rust, where all the inoculum that infects apples comes from galls on red cedars (junipers), and all the inoculum that infects red cedars comes from apple leaves and fruit. The epidemic on apples occurs during a four- to six-week period of basidiospore production in the spring. A second monocyclic epidemic on red cedars occurs during a brief period of aeciospore production in late summer.

POLYCYCLIC EPIDEMICS

In order for an epidemic to be considered polycyclic, there must be repeated complete infection cycles, that is, infection followed by pathogen development, new inoculum production, dispersal to new susceptible sites, and new infections, all within a single crop cycle. A good example is potato late blight, where a single cycle of infection, lesion development, sporulation, sporangium dispersal, and new infection can occur in as little as five days, and many overlapping cycles occur simultaneously during periods of favorable weather.

Each cycle can produce more than a ten-fold increase in numbers of sporangia landing on susceptible sites, and an explosive epidemic results. Cereal rusts behave similarly; a single urediniospore can infect to produce a pustule from which hundreds of new urediniospores can be released to infect and produce hundreds of new pustules repeatedly throughout the season. Most plant diseases caused by bacteria are polycyclic, and many plant viruses, with the aid of their insect vectors, also can produce repeated cycles of infection in one season.

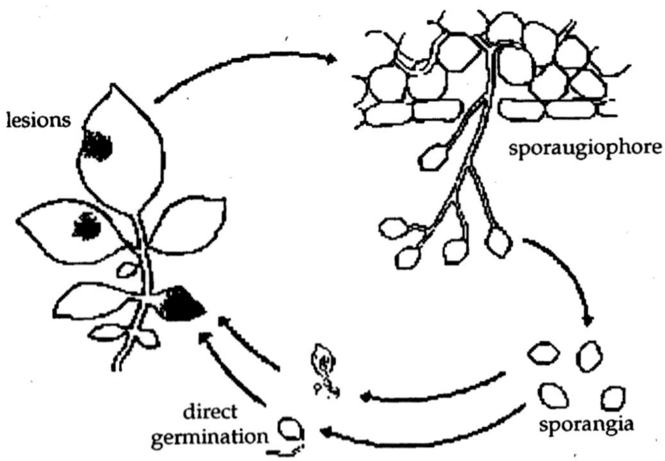

Late Blight Disease Cycle

Combinations of Monocyclic and Polycyclic Epidemics

Not every plant disease epidemic is clearly either monocyclic or polycyclic. Epidemics produced by fungi with two spore stages can have elements of both, sometimes in distinct phases, and sometimes occurring simultaneously. For example, the fungus *Venturia inaequalis*, causal agent of apple scab, produces ascospores on the dead, infected leaves that have overwintered from the previous season. These ascospores are released over a period of six to eight weeks in the spring and infect the newly expanding apple leaves.

Since no new ascospores are produced until the following spring, this component of the epidemic might be considered monocyclic. However, each leaf lesion within about ten days produces a second type of spores, conidia, that also can infect newly expanding leaves. Thus, for the early part of the season, a polycyclic epidemic is superimposed on a monocyclic epidemic. Since lesions produced by conidia cannot be distinguished from those produced by ascospores, the net effect appears to be a rapidly growing polycyclic epidemic.

An epidemic of tan spot on wheat, caused by *Pyrenophora tritici-repentis*, also comprises both polycyclic and monocyclic elements. The initial inoculum occurs on infected seed, and the epidemic starts as these seeds germinate to yield seedlings bearing leaf lesions. Conidia are disseminated by wind and splashing rain in repeated cycles of leaf infection throughout the period of crop development (the polycyclic phase). By the time flowering begins, the inoculum for the second phase of the epidemic has built up on the upper leaves of the plant, principally the flag leaf.

The flowers and developing grains are susceptible for a relatively brief period to infection by the inoculum extant at the time of flowering, and the lesions on the glumes of the developing grains do not produce inoculum capable of infecting more grains. Seed infection, therefore, is a monocyclic phenomenon. It is these infected seeds that provide a major means of carryover of this pathogen from one crop to the next and the initial inoculum for the new crop.

REFERENCES

Britton K.O, Redlin S.C., "Damping-off of flowering dogwood seedlings caused by Colletotrichum acutatum and Fusarium oxysporum", *Plant Disease,* 1995.

Canfield M.L, Baca S, Moore L.W., "Isolation of Pseudomonas syringae from 40 cultivars of diseased woody plants with tip dieback in Pacific Northwest nurseries", *Plant Disease,* 1986.

McRitchie J.J., "The felt fungus, Septobasidium", *Plant Pathology Circular Gainesville,* 1991.

3

Physiological Plant Diseases

Physiological plant disorders are caused by non-pathological disorders such as poor light, weather damage, water-logging or a lack of nutrients, and affect the functioning of the plant system. Physiological disorder are distinguished from plant diseases caused by pathogens, such as a virus or fungus. Whilst the symptoms of physiological disorders may appear disease-like, they can usually be prevented by altering environmental conditions. However, once a plant shows symptoms of nutrient deficiency it is likely that that season's yields will be reduced.

Causes of physiological disorders can be identified by examining:

— Where symptoms first appear on a plant- on new leaves, old leaves or all over?

— The pattern of any discolouration or yellowing- is it all over, between the veins or around the edges? If only the veins are yellow deficiency is probably not involved.

— Note general patterns rather than looking at individual plants- are the symptoms distributed throughout a group of plants of the same type growing together. In the case of a deficiency all of the plants should be similarly effected, although distribution will depend on past treatments applied to the soil.

— Soil analysis, such as determining pH, can help to confirm the presence of physiological disorders. Recent conditions, such as heavy rains, dry spells, frosts, etc, may also help to determine the cause of plant disorders.

WEATHER DAMAGE

Frost and cold are major causes of crop damage to tender plants, although hardy plants can also suffer if new growth is exposed to a hard frost following a period of warm weather. Symptoms will often appear overnight, affecting many types of plants. Leaves and stems may turn black, and buds and flowers may be discoloured, and frosted blooms may not produce fruit. Many annual plants, or plants grown in frost free areas, can suffer from damage when the air temperature drops below 40° Fahrenheit. Tropical plants may begin to experience cold damage when the temperature is 42-48° Fahrenheit, symptoms include wilting of the top of the stems and/or leaves, and blackening or softening of the plant tissue.

Frost or cold damage can be avoided by ensuring that tender plants are properly hardened before planting, and that they are not planted too early in the season, before the risk of frost has passed. Avoid planting susceptible plants in frost pockets, or where they will receive early morning sun. Protect young buds and bloom with horticultural fleece if frost is forecast. Cold, drying easterly winds can also severely inhibit spring growth even without an actual frost, thus adequate shelter or the use of windbreaks is important.

Drought can cause plants to suffer from water stress and wilt. Adequate irrigation is required during prolonged hot, dry periods. Rather than shallow daily watering, during a drought water should be directed towards the roots, ensuring that the soil is thoroughly soaked two or three times a week. Mulches also help preserve soil moisture and keep roots cool.

Heavy rains, particularly after prolonged dry periods, can also cause roots to split, onion saddleback (splitting at the base), tomatoes split and potatoes to become deformed or hollow. Using mulches or adding organic matter such as leaf mold, compost or well rotted manure to the will help to act as a 'buffer' between sudden changes in conditions. Water-logging can occur on poorly drained soils, particularly following heavy rains. Plants can become yellow and stunted, and will tend to be more prone to drought and diseases. Improving drainage will help to alleviate this problem.

Hail can cause damage to soft skinned fruits, and may also allow brown rot or other fungi to penetrate the plant. Brown spot markings or lines on one side of a mature apple are indicative of a spring hailstorm.

Plants affected by salt stress are unable to take water from soil, due to an osmotic imbalance between soil and plant.

NUTRIENT DEFICIENCIES

Poor growth and a variety of complaints such as leaf discolouration can be caused by a lack of plant foods. This may be due to shortages of necessary nutrients, or because the nutrients are present but not available to the plant. The latter can be caused by incorrect pH, shortages of water or an excess of another nutrient. Generally, the key to avoiding nutrient deficiencies is to ensure that the soil is healthy and contains plenty of well rotted organic matter rather than by feeding or treating individual plants.

Nutrient (or mineral) deficiences include:

— Boron deficiency

— Calcium deficiency

— Iron deficiency
— Magnesium deficiency
— Manganese deficiency
— Nitrogen deficiency
— Phosphorus deficiency
— Potassium deficiency

Boron Deficiency (plant disorder)

Boron (B) deficiency is a rare disorder affecting plants growing above a granite bedrock, which is low in boron. Boron may be present but locked up in soils with a high pH, and the deficiency may be worse in wet seasons.

Symptoms include dying growing tips and bushy stunted growth. Crop-specific symptoms include;

— *Beetroot*- rough, cankered patches on roots, internal brown rot.
— *Cabbage*- distorted leaves, hollow areas in stems.
— *Cauliflower*- poor development of curds, and brown patches. Stems, leafstalks and midribs roughened.
— *Celery*- leaf stalks develop cracks on the upper surface, inner tissue is reddish brown.
— *Pears*- new shoots die back in spring, fruits develop hard brown flecks in the skin.
— *Strawberries*- Stunted growth, foliage small, yellow and puckered at tips. Fruits are small and pale.
— *Swede (rutabaga)* and *turnip*- brown or grey concentric rings develop inside the roots.
— *Arecaceae* (*Palm Tree*) - brown spots on fronds & lower productivity.

Boron deficiency can be avoided by improving the moisture retaining capacity of light soils, and ensuring pH is kept below 7. Borax can be raked into the soil at 35 g/m^2 (1 oz/yd^2).

Calcium Deficiency

Calcium (Ca) deficiency is a plant disorder that can be caused by insufficient calcium in the growing medium, but is more frequently a product of a compromised nutrient mobility system in the plant. This may be due to water shortages, which slow the transportation of calcium to the plant, or can be caused by excessive usage of potassium or nitrogen fertilisers.[citation needed]

Symptoms

Calcium deficiency symptoms appear initially as generally stunted plant growth, necrotic leaf margins on young leaves or curling of the leaves, and eventual death of terminal buds and root tips. Generally the new growth of the plant is affected first. The mature leaves may be affected if the problem persists. Crop-specific symptoms include:

Apple : 'Bitter pit' – fruit skins develop pits, brown patches appear in flesh and taste becomes bitter. Can occur when fruit is in storage. Bramley apples are particularly susceptible.

Cabbage and Brussels sprouts : Internal browning.

Carrot: 'Cavity spot' – oval spots develop into craters which may be invaded by other disease causing organisms.

Celery : Stunted growth, central leaves stunted.

Tomatoes and peppers : 'Blossom end rot' – Symptoms start as sunken, dry decaying areas at the blossom end of the fruit, furthest away from the stem, not all fruit on a truss is necessarily affected. Sometimes rapid growth from high-nitrogen fertilisers may cause blossom end rot.

Treatment

Calcium deficiency can be rectified by adding Agricultural lime to acid soils, aiming at a pH of 6.5, unless the plant in question specifically prefers acidic soil. Organic matter should be added to the soil in order to improve its moisture-retaining capacity.

Plant damage is difficult to reverse, so take corrective action immediately. Make supplemental applications of calcium nitrate at 200 ppm nitrogen. Test and correct the pH if needed because calcium deficiency is often associated with low pH.

Iron Deficiency

Iron (Fe) deficiency is a plant disorder also known as 'lime-induced chlorosis'. A deficiency in the soil is rare. Iron can be unavailable if pH is too high or if the soil is waterlogged, or has been overfertilised with phosphorus. Can be confused with manganese deficiency.

Any plants may be affected, but raspberries and pears are particularly susceptible, as well as most acid-loving plants such as azaleas and camellias.

Symptoms include leaves turning yellow or brown in the margins between the veins which may remain green, while young leaves may appear to be bleached. Fruit is of poor quality and quantity.

Iron deficiency can be avoided by choosing appropriate soil for the growing conditions (e.g., avoid growing acid loving plants on lime soils), or by adding well-rotted manure or compost. Iron deficiency can be caused by acidic conditions.

Magnesium Deficiency

Magnesium (Mg) deficiency is a plant disorder with two main causes. Magnesium can be easily washed out of light soils in wet seasons or excessive potassium fertiliser usage can cause also Mg to become unavailable to the growing plant.

This disorder particularly affects potatoes, tomatoes, apples, currants and gooseberries, and chrysanthemums.

Symptoms include, yellowing between leaf veins, which stay green, giving a marbled appearance. This begins with older leaves and spreads to younger growth. Can be confused with virus, or natural aging in the case of tomato plants. Fruits are small and woody.

Mg deficiency can be rectified in the short term by applying a foliar feed fortnightly, with epsom salts diluted at a rate of 200g per 10 litres of water (8oz per 2½ gal). In the longer term add dolomitic limestone if soil pH allows, or other Mg containing rocks such as Kieserite. Reduce usage of potash fertilisers if this may be contributing to the problem

Manganese Deficiency

Manganese (Mn) deficiency is a plant disorder that is often confused with, and occurs with, iron deficiency. Most common in poorly drained soils, also where organic matter levels are high. Manganese may be unavailable to plants where pH is high.

Affected plants include onion, apple, peas, French beans, cherry and raspberry, and symptoms include yellowing of leaves with smallest leaf veins remaining green to produce a 'chequered' effect. The plant may seem to grow away from the problem so that younger leaves may appear to be unaffected. Brown spots may appear on leaf surfaces, and severely affected leaves turn brown and wither.

Prevention can be achieved by improving soil structure. Do not over-lime.

Nitrogen Deficiency

Nitrogen (N) deficiency in plants can occur when woody material such as sawdust is added to the soil. Soil organisms will utilise any nitrogen in order to break this down, thus making it temporarily unavailable to growing plants. 'Nitrogen robbery' is more likely on light soils and those low in organic matter content, although all soils are susceptible. Cold weather, especially early in the season, can also cause a temporary shortage.

All vegetables apart from nitrogen fixing legumes are prone to this disorder. Symptoms include poor plant growth, leaves are pale green or yellow in the case of brassicas. Lower leaves show symptoms first. Leaves in this state are said to be etiolated with reduced chlorophyll. Flowering and fruiting may be delayed.

Prevention and control of nitrogen deficiency can be achieved in the short term by using grass mowings as a mulch, or foliar feeding with manure, and in the longer term by building up levels of organic matter in the soil. Sowing green manure crops such as grazing rye to cover soil over the winter will help to prevent nitrogen leaching, while leguminous green manures such as winter tares will fix additional nitrogen from the atmosphere.

Phosphorus Deficiency

Phosphorus (P) deficiency is a plant disorder that is most common in areas of high rainfall, especially on acid, clay or poor chalk soils. Cold weather can cause a temporary deficiency.

All plants may be affected, although this is an uncommon disorder. Particularly susceptible are carrots, lettuce, spinach, apples, currants and gooseberries. Symptoms include poor growth, and leaves that turn blue/green but not yellow—oldest leaves are affected first. Fruits are small and acid tasting. Phosphorus deficiency may be confused with nitrogen deficiency.

It can be controlled by applying organic sources of phosphorus such as rock phosphate.

Plants that are naturally adapted to low levels of available soil phosphorus, however, are more likely to suffer from phosphate poisoning: the key is to provide the right level for any particular plant type, neither too high nor too low.

Potassium Deficiency

Potassium deficiency, also known as potash deficiency, is a plant disorder that is most common on light, sandy soils, as well as chalky or peaty soils with a low clay content. It is also found on heavy clays with a poor structure.

Plants require potassium ions (K^+) for protein synthesis and for the opening and closing of stomata, which is regulated by proton pumps to make surrounding guard cells either turgid or flaccid. A deficiency of potassium ions can impair a plant's ability to maintain these processes.

The deficiency most commonly affects fruits and vegetables, notably potatoes, tomatoes, apples, currants, and gooseberries, and typical symptoms are brown scorching and curling of leaf tips, and yellowing of leaf veins. Purple spots may also appear on the leaf undersides.

Deficient plants may be more prone to frost damage and disease, and their symptoms can often be confused with wind scorch or drought.

Prevention and cure can be achieved in the shorter term by feeding with home-made comfrey liquid, adding seaweed meal, composted bracken or other organic potassium-rich fertilisers. In the longer term the soil structure should be improved by adding plenty of well rotted compost or manure. Wood ash has high potassium content, but should be composted first as it is in a highly soluble form.

REFERENCES

Agrios, George N., *Plant Pathology*, Fifth ed. Amsterdam: Elsevier, 2005.

Bradbury, J. F., *Guide to Plant Pathogenic Bacteria*, Farnham Royal, England: CAB International, 1986.

4

Diseases of Vegetable Crops

BEAN (SNAP, LIMA, AND BUTTER BEANS)

Phaseolus spp

Bacterial Blights: Halo Blight (bacterium - *Pseudomonas syringae* pv. *phaseolicola*); Common Blight (bacterium - *Xanthomonas campestris* pv. *phaseoli*): Plants infected with the halo blight bacterium form greenish-yellow circles around each lesion. Interior of the lesion turns brown. With age, lesions enlarge and coalesce. The entire leaf finally drops. Stem lesions appear as long, reddish colored spots. When the plant begins to set fruit, lesions are formed at the nodes which girdle the stem. This reduces fruit development. Common blight-infected pods do not exhibit the greenish-yellow halo around the lesion like halo blight lesions. Infected leaves with halo blight turn yellow and slowly die while those with common blight turn brown and drop quickly. Both organisms are seed-borne. Entry into the plant is through the leaf stomata. Rain and damp weather encourage development of these diseases. Common blight is more of a problem in warm weather while halo blight is favored by cool temperatures. Both bacteria can live in the soil for two years on plant residue. To control bacterial blight of beans, seed grown in the western United States should be planted. Avoid spreading the disease by not entering the field when the foliage is wet. Follow a three year rotation.

Anthracnose (fungus - Colletotrichum lindemuthianum): This is a seed-borne fungus which attacks all above ground portions of the plant. Infected seed are marked by dark, sunken lesions that extend through the seed coat. Stem lesions are oval and sunken. The center of the lesion is dark brown with purplish to red borders. In early stages, the fungus develops along the veins and becomes purplish to red in color. In advanced stages, leaves become ragged. Infection of the pods results in small, reddish, elongated spots. Older spots are sunken and have brown to reddish-brown borders. The disease is favored by cool, wet springs and falls. It disappears during hot, dry summers. The fungus can survive in the soil for two years in plant debris. Control is obtained by: (1) the use of disease-free seed, (2) crop rotation, (3) not entering fields when plants are wet, and (4) spraying with fungicides.

Cercospora Leaf Spot (fungi - Cercospora spp.): Lower foliage becomes marked by irregular tan spots one-eighth to one-fourth inch in diameter. Severe infection causes defoliation and plant stunting. It is reported to attack the pod but has only been observed on foliage in Texas. Infection is most severe during periods of extended rainfall, high humidity and temperatures between 75 to 85 degrees F. No resistance exists among varieties. Fungicide sprays should begin at first sign of disease and continued during cool, rainy conditions.

Root Rot (fungus - Rhizoctonia solani): Bean seed may rot in soil or the young seedling may become stunted. A reddish-brown canker is formed on the stem. Cankers may completely girdle the stem or may only partially girdle it, causing severe stunting. Beans should be planted after the soil has warmed to above 69 degreesF. Beans should follow a grass type crop.

Fusarium Root Rot (fungus - Fusarium solani f. sp. phaseoli): Plants infected with Fusarium are characterised by a reddish discoloration of the tap root. Affected plants are stunted with yellow leaves. Young rootlets formed in the area of the lesion are killed. If weather conditions are favorable, a normal crop may be produced. Avoid soils where Fusarium has been a problem. Long rotations (four to five years) will help reduce losses.

Pod Blight (fungus - Diaporthe phaseolorum): Pod blight of lima beans is first observed as brown pustules of irregular shape on the leaves. Lesions grow to one-fourth to three-fourths of an inch in diameter. During the latter part of the growing season, the fungus spreads to nearby pods, where it causes a pale watery spot. The spot enlarges and becomes darker with age. On pods the spot is marked by dark brown to black pustules on the surface arranged in a ring.

Use seed grown in the western United States and use a three- to four-year rotation. Follow a fungicide program to control the disease when it occurs consistently.

Mosaic (virus): Leaves become puckered and mottled with light and dark-green areas. Infected plants become stunted. The virus is seed-borne and can be spread by aphids. Losses can be reduced by growing resistant varieties and following an approved aphid control program.

Curly Top (virus): Infected plants are stunted and have distorted foliage. It is spread by the beet leafhopper. Use resistant varieties and practice good insect control.

Sun Scald: In the early spring beans are often affected by a condition in which the young leaves turn light tan in color and die. This may happen to the entire leaf or to only a portion of the leaf. The conditions favoring disease development are cool to moderate temperatures, extended periods of high humidity, and cloudy days followed by a bright sunny day. Damage is usually restricted to only a few scattered leaves.

Rust (fungus - Uromyces phaseoli): Small reddish-brown pustules form on lower side of leaves. The fungus overwinters in crop residue. If rust has been severe, a rotation program should be practiced. Resistant varieties should be used when past experience indicates rust to be a problem. Apply approved fungicides at first sign of infection in the fall.

Powdery Mildew (fungus - Erysiphe polygoni): Powdery mildew is characterised by a white powdery growth on the foliage. Infected pods and foliage become malformed. The fungal spores are spread by wind. Spray with approved fungicides. Powdery mildew seldom becomes an economic problem.

Watery Soft Rot (fungus - Sclerotinia sclerotiorum): The fungus affects the stems, leaves and pods of beans. First signs of infection are small, soft, watery spots that enlarge rapidly under cool, moist conditions. They may enlarge and coalesce and the stem is girdled. Infected pods turn into a soft, watery mass. Following the watery stage the affected tissues dry out and turn brown. Within a short time the brown areas are covered with a dense white fungal growth. With age, the white fungal growth turns gray and is dotted with small, hard black bodies called sclerotia. Most losses occur in shipping. Infected beans tend to stick together.

The disease is favored by temperatures ranging between 60 and 70 degrees F. Long periods of high humidity also favor the development of white mold. Large plants with heavy vine growth encourage disease development.

Sclerotia fall to the ground at maturity where they can lay dormant for as long as ten years. When weather conditions are again favorable, the sclerotia begin growth again. The fungus enters beans directly where pods and leaves come in contact with the developing fungus.

Sclerotia produce small mushroom-like structures which contain thousands of ascospores. The spores are ejected into the air landing on plant parts such as blossoms or decaying leaves and begin to develop. This may be repeated many times resulting in widespread infection.

Every effort should be made to improve air circulation between plants and rows. This can be done by increasing row spacing and decreasing the seeding rate. Excessive applications of nitrogen favor heavy vine growth and should be avoided. During periods of extended cool temperatures and high humidity, fungicides should be applied on a preventive schedule.

Baldhead: Beans emerge and produce cotyledons but nothing is formed above the cotyledonary leaves. This can result from injury from soil insects or fungi, and results from mechanically damaged seed.

BEET (TABLE)

Beta vulgaris

Cercospora Leaf Spot (fungus - *Cercospora beticola*): Cercospora leaf spot causes small, brown lesions with reddish-purple borders. As the spots enlarge, the center becomes gray with purple borders. Severe infection causes the leaves to drop. Crop residue should be plowed under. High humidity and temperatures around 75 to 80 degrees F are optimum for fungal growth. Foliar sprays with a preventative fungicide should be applied when symptoms first appear.

Curly Top (virus): Plants infected with curly top die prematurely. Young leaves roll inward and eventually turn yellow and die.

Mosaic (virus): Leaves appear puckered and mottled with light and dark green areas. Plants are stunted. Chlorosis occurs on infected leaves.

Powdery Mildew (fungus - Erysiphe polygoni): This is a relatively new disease of both sugar and table beets.

Infected leaves are covered by a white, powdery fungal growth. Affected leaves turn red. Sulfur-containing fungicides are recommended.

CABBAGE, CAULIFLOWER, BRUSSEL SPROUTS, AND BROCCOLI

Brassica spp.

Black Rot (bacterium - Xanthomonas campestris pv. campestris): Infection usually occurs near margins of leaves. Yellow areas develop along the margins, progressing into the leaf in an inverted-V shape. Veins in the affected areas are black. The bacterium may attack cruciferous plants at all stages of growth. A black discoloration is observed in the stem when it is split lengthwise. Cauliflower is very susceptible to black rot and the interior of the stem may be destroyed by secondary soft rot organisms that attack the black rot infection. It is a seed-borne bacterium which can persist in the soil for only a short time. Some wild plants may act as hosts harboring the bacterium while crucifers are absent from commercial fields. Young plants resulting from infected seed serve as a source of secondary infection. Under crowded conditions in the field and during periods of rain, the bacteria spread quickly to nearby healthy plants. Control is achieved by following a 2 year rotation, by planting disease-free seed, and seed that has been treated with hot water to eliminate seed-borne bacteria.

Fusarium Yellow (fungus - Fusarium oxysporum f. sp. conglutinans): The first indication of the disease is the yellowish-green color of the foliage. Plants appear wilted and stunted. Infected plants usually show a curved mid-rib and the leaf grows on only one side. The disease may be easily confused with black rot since many symptoms are similar. The fungus can live in the soil for a number of years without a host plant. It infects plants through young rootlets and wounds caused by transplanting or insects. It develops in the water conducting vessels, causing a brown discoloration similar to that produced by the black rot bacterium. The disease is checked by very cold or very hot soil tempera- ture. Resistant varieties and long rotations will avoid crop loss.

Downy Mildew (fungus - Peronospora parasitica): The fungus is more noticeable on the underside of leaves as a gray, fluffy, downy growth in well defined spots. Outer leaves of the head will develop small black spots when infected with the mildew fungus. The fungus overwinters in crop refuse. Sanitation, rotation and spraying with a protective fungicide are the most effective control methods. During periods of high humidity and cool temperature, the fungus is difficult to control and is almost always present in cabbage fields.

Black Leg (fungus - Phoma lingam): First symptoms occur on leaves and stems as small spots with ashen-gray centers and black dots. The stem lesions gradually enlarge extending to the roots. In advanced stages the root system develops a dry rot. Wilted leaves tend to remain attached to the stem instead of defoliating like plants infected with yellows or black rot. The fungus is carried on seed and can persist on infected plant residue for one or two years. Spread is dependent on dew, rain or irrigation water. Avoid soils where cabbage or related plants were grown within the last four years. Destroy crop residues by shredding and deep burial of crop residue.

Powdery Mildew (fungus - Erysiphe polygoni): This disease occurs infrequently on cabbage and related plants. A white, powdery growth is observed on the upper surface of leaves. Protective fungicides can be applied if plants become severely infected.

Alternaria Leaf Spot (fungus - Alternaria brassicae): The first symptom is a minute dark spot on seedling stems and on the leaves. These spots enlarge and are marked with concentric rings, giving a bull's eye appearance. The fungus overwinters on cabbage residue or on seed. Spores are disseminated by wind or water. Hot water treatment, as recommended for black rot, will rid the seed of this organism. Fungicide application will prevent the fungus from developing in the field.

Rhizoctonia or Wire Stem Disease (fungus - Rhizoctonia solani): This disease may appear at different stages of growth. A damping-off phase of the disease will cause young seedlings to die. After seedlings are older, they may be attacked but lesions seldom completely girdle the stems. The infected stems are somewhat smaller than normal and are tough and woody. Older plants can also be attached causing a head rot or a root rot. The causal agent is a common soilborne fungus that attacks many plants. Crop rotation and planting healthy transplants are the most effective means of control.

Internal Tip Burn (physiological): Tip burn causes leaf margins to turn brown. The leaves are buried in the head. Exact nature of the problem has been associated with poor water movement within the plant.

Soft Rot (bacterium - Erwinia carotovora): Soft rot also occurs most commonly when fields become water saturated. Stems become decayed and have a foul odor. During storage and transit, a slimy decay with a foul odor develops. The disease begins in areas that have been bruised, particularly during periods of high temperature and humidity prevail. Proper handling during harvesting, packaging and storing along with keeping temperatures low are the best means of control. Follow a long rotation and plant on raised beds in well drained soil to prevent field infections.

Cabbage Mosaic (virus): Leaves display green and yellow mottled areas. Leaf veins may be lighter in color. Keeping fields and surrounding areas free of weed hosts will help reduce the disease.

CANTALOUPE, CUCUMBER, SQUASH PUMPKIN AND CRENSHAW

***Cucumis and Cucurbita* spp.**

Downy Mildew (fungus - Pseudoperonospora cubensis): Angular yellow spots appear on the upper surface of the leaf during periods of high humidity. The underside of leaves, opposite the yellow spots, becomes covered by a grayish growth which is the spore producing structures of the fungus. This growth is more noticeable early in the morning when heavy dew is present. Spores are easily carried by wind from diseased plants. Cool temperatures along with free moisture are ideal for mildew infection and spread. Hot, dry weather may reduce or stop disease development. Resistant varieties should be used when possible. Preventative fungicide applications should begin at the time of blooming and continued at 7 to 10 day intervals until the crop is harvested.

Powdery Mildew (fungus - Erysiphe cichoracearum and Sphaerotheca fuliginea): This disease appears on the leaves as a white powdery mass composed of the spore-bearing structures of the fungus. Severely infected leaves shrivel and die. The fungus can also grow on petioles and young stems. Infected plants are yellow, stunted, and may die. Fruits are not attacked, but are usually small and deformed. Powdery mildew is favored by cool, dry weather. Spores are air-borne. Varieties of most cucurbits are available that show some degree of resistance to powdery mildew. Preventative fungicide applications will effectively control powdery mildew. New races of the fungus have appeared which are resistant to some fungicides. Several applications at 7 to 10 day intervals may be necessary for optimum control.

Anthracnose (fungus - Colletotrichum lagenarium): First symptoms of this disease are spots on the foliage that begin as yellowish or watersoaked area. Spots enlarge and turn brown to black in color. Diseased tissue dries and the center of the spots fall out, giving the leaf a "shot-hole" appearance. Symptoms on fruits consist of circular, black, sunken, cankers varying in size depending on the host plant. When abundant moisture is present, the center of the spot is surrounded by a gelatinous pink mass. The fungus overwinters on old cucurbit vines and residues. It may also be seed-borne. Plants may be infected at any stage of growth. Disease appearance in the field depends mainly on rainy, cool weather for a period of several days. All above ground portions may be infected. Seed treatment, crop rotation and destruction of crop residues are important practices for the control of this disease. Preventative fungicide applications, as recommended for downy mildew control, should be made at seven to 10 day intervals, or more often, if weather conditions are favorable for disease development. Preventative fungicide applications will protect plants against infection.

Gummy Stem Blight (fungus - Didymella bryoniae): This disease affects leaves and stems, and is more prominent in the crown at soil level. Leaf symptoms begin with irregular spots that dry and drop out giving the lesion a ragged appearance. The most conspicuous phase of the disease is the brown exudation in the crown of infected plants. Vine cankers are found near the soil line, producing a gummy brown ooze. The fungus overwinters in the soil and on crop residue. The fungus is seed-borne. Once the fungus becomes established, millions of spores are produced which can be readily disseminated to other plants by rain, wind or mechanical equipment. Using disease-free seed or seed that has been treated with a fungicide will prevent the disease from becoming established in new fields. Crop rotation should be practiced. Spraying with some fungicides has also shown to be effective.

Angular Leaf Spot (bacterium - Pseudomonas syringae pv. lacrymans): This disease appears on leaves, stems and fruit. Spots are small, angular, straw-colored and watersoaked.

Leaf spots often dry and fall out giving the leaf a "shot-hole" appearance, similar to those caused by anthracnose. Spots on fruit are usually smaller and circular in shape. Bacteria overwinter in crop residue and on seed. Hard rains splash the bacteria to stems and leaves. The disease may reach epidemic proportions during periods of heavy rains, particularly if temperatures remain high. Since the disease is seed-borne, the use of disease-free seed treated with a fungicide is recommended. Crop rotation with non-cucurbit crops is also helpful for control of angular leaf spot. Application of copper fungicide will assist in control. Cucurbits, however, are sensitive to copper when young, and repeated applications of copper may cause yellowing of foliage around the edges of the leaves.

Alternaria Leaf Spot (fungus - Alternaria cucumerina): Small, circular watersoaked spots first appear on the leaves, later enlarging to one and one-half inch or more in diameter. Definite concentric rings and margins appear giving the diseased area a "bull's eye" appearance. The fungus overwinters in crop residue and on other cucurbit plants. Numerous air-borne spores are produced on diseased spots. Crop rotation and seed treatment help. A preventative fungicide program should be followed when the disease occurs consistently.

Choanephora Wet-Rot (fungus - Choanephora cucurbitarum): This disease occurs almost exclusively on crenshaw, squash and pumpkins. The blossom ends of fruit turn black, watersoaked, and covered by a fungal growth. The fungus is usually confined to the end of the fruit, but if conditions favorable for disease development prevail, the entire fruit may decay. Greatest damage by this disease is caused during prolonged damp weather.

Fusarium Rind Rot (fungus - Fusarium roseum): Rind rot of cantaloupes caused by Fusarium roseum, caused considerable damage in the Rio Grande Valley in recent years. Rots appear on the melon rind which at first may go unnoticed. However, after peeling the fruit, large brown decomposed areas may be found in the flesh. Since the disease may go unnoticed in the field, melons may be harvested, packed and shipped without the handlers being aware of the problem.

Cucumber Mosaic (virus): This virus causes prominent vein clearing in young leaves, followed by pronounced mosaic mottle on older leaves, and stunting. With cucumber, the fruit is prominently mottled, misshapen and blistered. The virus is spread by aphids. It does not affect watermelon.

Squash Mosaic (virus): This virus does not affect cucumber, cantaloupe, honeydew and watermelon fruit. Symptoms on leaves are faint mottling and distortion. On squash, prominent vein clearing and mottling occurs. Yellow variety fruits are green with isolated yellow zones and misshapen fruit. The virus is spread by the cucumber beetle.

Tobacco Ringspot Virus (virus): Chlorotic stippling, ringspots and mild vein clearing occur on leaves on cantaloupe, honeydew, and cucumber. Squash leaves are severely affected, often showing leaf perforations. Fruit of yellow varieties are severely misshapen with irregular sunken or raised dark green spots. Watermelon plants have shortened internodes and pronounced upright growth of terminals. On fruits, concentric rings and pimples are scattered over the green rind.

Watermelon Mosaic Virus (virus): Vein clearing and mosaic patterns occur on cucumber, cantaloupe, and squash. Older plants are severely stunted and leaves are malformed with dark green blisters. Runners are twisted and stand upright, a symptom expression usually referred to as "snake head." Fruits are small. Squash fruit develop a typical mosaic pattern. Yellow squash are green with yellow spots. Infected zucchini squash are rough and stunted. In cucumbers, the fruit is mottled partially at the stem end. The virus is spread by aphids and possibly other sucking insects. No resistance exists among currently used varieties. Early insect control is important for the prevention of this disease.

Charcoal Rot (fungus - Macrophomina phaseolina): Symptoms resemble those of gummy-stem blight and other vine diseases of melons. First symptoms are dying of leaves close to the crown. As the disease progresses, entire runners wilt and die. Close observation of the crown will reveal brown cankers on the stems both below and above the soil line. Black, hard bodies produced by the fungus, known as sclerotia, are found on the surface of the cankers. Crop rotation may be of limited use since the fungus affects a large number of hosts, including corn and grain sorghum. There appears to be some difference among the varieties. Soil fumigation with either liquid or gas compounds kept for a few days under a plastic cover may offer another means of control.

Fusarium Wilt (fungus - Fusarium oxysporum f. sp. melonis): This is one of the most severe diseases, especially of watermelons, that causes sudden wilting and plant death about the time that fruits are beginning to enlarge. It also attacks the plant at other growth stages especially during the seedling stage when it may produce damping-off. The causal fungus is soil and seed-borne. The disease seems to be most severe when infected seed are used. Serious losses may occur even where watermelons have not been previously grown. Use disease free seed and utilise long rotational programs. Some varieties are more resistant than others.

CARROTS

Aster Yellows Symptoms

Aster Yellows (Mycoplasma): The first symptom of aster yellows is yellowing of the foliage, followed by excessive growth and bunching of shoots. Older leaves become twisted and may fall off. Leaves later have a bronzed appearance. Roots are misshapen, and are of poor quality with a proliferation of adventitious roots. This organism is transmitted by leafhoppers. weeds in and surrounding the crop can serve as a reservoir for this pathogen. The disease usually occurs sporadically with little economic loss.

Damping-off Symptoms

Damping-off (fungi- Rhizoctonia, Fusarium, etc.): Like many other vegetables, carrot seedlings are susceptible to several species of soilborne fungi, particularly during the periods of cool, wet weather. Infected seedlings wilt, turn brown and die, resulting in poor stands. To prevent damping-off, seeds should be planted on a bed to allow for drainage. Seedlings should not be overwatered.

Leaf Spot Symptoms

Leaf Spot (fungus - Cercospora carotae): This disease may occur at any position on the leaf but is most common on margins. Spots are circular in shape and, with age, they coalesce to form larger spots. Lesions on the leaves are, sometimes surrounded by a lighter-colored halo. Lesions on the petioles are enlongated with a pale center and dark margin. Spots on the petiole may encircle it causing defoliation. Leaf spot is distinguished from leaf blight by the nearly circular, sharply defined lesions with a yellow halo. The disease usually occurs during the latter part of the growing season and can be controlled with the same fungicide program as used for leaf blight.

Southern Blight Symptoms

Southern Blight (fungus - Sclerotium rolfsii): This disease occurs close to harvest and is associated with warm temperatures that follow heavy rains. Plants die in clusters within a field. A bright, white mycelium is found on the surface of the soil, along with spherical, tan resting bodies of the fungus that resemble a mustard seed. By the time the fungus is seen on the surface of the soil, the carrot is rotted. There are no control recommendations. Harvesting should be done as quickly as possible in affected fields. Harvested carrots should be processed and cooled as quickly as possible to minimise post- harvest decay caused by the fungus.

Black Root Rot Symptoms

Black Root Rot (fungus - Thielaviopsis basicola): This is primarily a post-harvest problem, although soil is the source of the pathogen. Black, irregular areas are seen on the roots. Physical injury predisposes carrots to infection, but disease can be minimised by washing soil from them, cooling them as quickly as possible to at least 45°, and by rinsing them in chlorinated water before placing them in bags.

Growth Crack Symptoms

Growth Crack (physiological): The length of the carrot is split. This is caused mainly by fluctuating soil moisture throughout the growing season. If there is a high incidence in a field, then carrots should be watered more regularly to prevent wide moisture fluctuations.

Powdery Mildew Symptoms

Powdery Mildew (fungus - Erysiphe polygoni): This disease is not usually serious. The surface of affected leaves are covered by a white mass of the fungus, which has a powdery texture. Symptoms may also be found on petioles. Fungicide applications at 10 to 14 day intervals will control the disease.

Cotton Root Rot Symptoms

Cotton Root Rot (fungus - Phymatotrichopsis omnivora): Carrots are usually grown during cooler times of the year when this fungus is not active and so, they escape this disease. In geographic areas where this pathogen is present, the disease can be seen later in the spring on carrots growing in lighter soils, particularly following an irrigation. Areas of dead and wilted plants are seen in a field. These resemble the dead areas that can also be caused by the southern blight fungus, but there is no fungal growth on the surface of the soil . When carrots are pulled out, they are not rotted and a sheath of soil clings to them. When this soil is removed, infection cushions and mycelial strands of the fungus can be seen on the surface of the carrot. A microscopic examination for the the characteristic cruciform hyphal branching will confirm the diagnosis. Planning for an earlier harvest will prevent disease losses.

Leaf Blight Symptoms

Leaf Blight (fungus - *Alternaria dauci*) Infection occurs mostly on older leaves, but younger leaves may also become infected. Leaf blight first appears as indefinite brown to black areas with pale yellow centers. Infected leaves shrivel when infection is heavy. Under these conditions foliage appears as if burned by fire. The fungus overwinters in infected crop refuse. Air-borne spores of the fungus are produced in large numbers on old lesions during periods of high humidity. Spores may also be carried on seed. The fungus requires moisture on leaves for infection. A preventative fungicide program should be followed where the disease is normally present, using a spray interval of 10 to 14 days.

Root Knot Nematode Symptoms

Root Knot Nematode (nematode - Meloidogyne spp.): There are no diagnostic symptoms seen on above-ground plant parts. When infected carrots are examinged, galls can be seen on adventitious roots and there is often multiple tap-root formation or malformed roots. Although deformed carrots are culled at the packing shed, nematodes contained in the roots are not harmful to humans when consumed. A soil test for nematodes can be done before planting to determine the potential for a problem. Pre-plant nematicides can be used where the soil is infested. Nematode populations in soil can also be reduced with a weed-free fallow or by rotation with cereals.

CELERY

Apium graveolens var. *dulce*

Cercospora Blight (fungus - Cercospora apii): Symptoms begin as small, yellowish spots on both sides of the foliage. Spots enlarge quickly changing from yellow to ash-gray with the tissue drying in affected areas. The fungus also attacks leaf petioles and stems. Under conditions of high humidity and warm weather, a gray velvety growth can be seen on the surface of affected spots. The fungus is seed-borne and spores can also be blown in air currents for long distances. Warm temperatures are required for fungus development. Control is obtained by using disease-free seed treated with fungicides and spraying plants with a protective fungicide at regular intervals when disease represents a threat.

Late Blight (fungi - Septoria apiicola): Small, yellowish spots develop on the underside of the leaves which later spread to the entire foliage. Spots enlarge and turn brown, similar to lesions caused by Cercospora. Minute, black fruiting bodies can be found scattered in the affected areas. Lesions with fruiting structures can be seen on petioles and stems. High humidity, free water, and cool temperatures favor disease development. Fungicides recommended for *Cercospora* blight will also control late blight.

Stalk Rot (fungi - Rhizoctonia solani): A serious disease of celery that is favored by high temperatures and moist conditions. Symptoms are first seen as small lesions on the base of the petioles, near the ground. As the disease progresses, the spots are enlarged, appear watersoaked at first, turning later to a brick-red color. When matured, the spots are sunken and turn to a dark brown color that is characteristic of the disease. The lesion can be a few or numerous, making it necessary to trim a lot of the leaves, thus reducing quality and yield. Shallow planting on raised beds and fungicide applications serve to reduce disease losses.

CORN (SWEET)

Zea mays var. *saccharata*

Seed Rots and Seedling Disease (fungi - Pythium spp. Macrophomina phaseolina, Gibberella zeae, Penicillium oxalicum and others): Both seed rots and seedling disease can cause poor stands. Cool wet soils slow seed germination and development of young seedlings so that there is exposure to fungi for a longer period of time. Low quality seed also produce seedlings that are weak and survive poorly in cold wet soils.

Control is obtained by using high quality seed which have been treated with a protective fungicide. Sweet corn should be planted on raised bed after the soil temperature is above 55 degrees F.

Stalk Rot and Kernel Rot (fungus - Fusarium spp.): Infected plants are stunted and delayed in maturity. During periods of high wind plants often lodge. Ears may hang downward on the stalk. The Gibberella stage of the fungus will infect kernels causing them to be pink in color. Infected ears have a strong odor and should not be used as food or feed.

Control stalk rots by rotating with non-related crops, planting in well drained soils and by using treated seed.

Southern Corn Leaf Blight (fungus - Bipolaris maydis): The disease is easy to recognise under field conditions. Spots on the leaves are tan to brown in color. On the ear the fungus causes oblong, bleached spots which penetrate through the shuck layers and finally into the ear.

The fungus overwinters in crop residue and produces spores which can be carried for long distances by wind.

Northern Corn Leaf Blight (fungus - Exserohilum turcicum): This disease is found in most sweet corn fields, yet is seldom severe enough to cause economic loss. Spots produced are larger than those caused by the southern corn leafspot fungus. Spots are from one to six inches long and one-half to one inch wide. With maturity, the center of the spot has a dark brown color. Infection occurs first on older foliage. High humidity and temperatures between 60 to 80 degrees F favor disease development. Varieties vary in their reaction to the fungus.

Brown Leaf Spot (fungus - Physoderma maydis): The fungus causing this disease occurs in most fields but seldom does economic damage. Infection requires high temperatures and presence of surface moisture. The first symptom of the disease is small circular spots. As they mature they turn dark brown. Rotation and deep burial of stalks will help reduce losses to this fungus.

Downy Mildew (fungus - Peronosclerospora sorghi): Infected plants are chlorotic, stunted and have striped leaves. Infected leaves have a downy growth on the underside, toward the basal part. Potential infection is increased when the crop is grown in soil previously grown to infected sorghum, field corn or sweet corn. Although high populations of spores are produced on the leaf surface, they are short lived and require extended periods of high humidity for infection. Overwintering spores produced between leaf veins exist in the soil for long periods. Practices which hasten the breakdown of crop residue will help reduce the amount of inoculum carried over in the soil. Varieties vary in their reaction to this disease. Growers should consult their county Extension agent for current hybrids and their reaction to this disease.

Crazy Top Downy Mildew (fungus - Sclerophthora macrospora): This disease is a problem when fields become flooded early in the life of the plant. The fungus produces swimming spores which require water for mobility.

Infected plants are sterile and have numerous shoots at the base of the stalk. Leaves are thickened, distorted, of a lighter green color than normal leaves. Tassels and ears develop green leafy shoots.

The fungus is commonly found in grasses along the edge of the field. Spores are washed into the field in flood water. Infection of the young corn plants takes place at this time.

Varieties vary in their reaction to this fungus. Due to the low percentage of occurrence, little has been done to rate varieties for their reaction. Avoid fields that flood regularly and plant on a raised bed which will help reduce the exposure of young seedlings to standing or flowing water.

Common Rust (fungus - Puccinia sorghi): Common rust occurs in most home gardens and commercial fields, but seldom causes economic losses. Infected leaves have raised spots or pustules formed primarily on the upper surface. The pustules are rectangular, brick red and occur in bands on the leaf. Spores are produced in the pustules, which are blown to neighboring leaves where infection can be repeated. Infection is encouraged by high humidity and cool temperatures (60 to 70 degrees F).

Common Smut (fungus - Ustilago maydis): Common smut is often found in fields of sweet corn. Losses from common smut vary based on the amount of infected plants in the field. Galls are formed as the common smut fungus causes cells of the corn plant to increase in size and number. These galls at first are covered with a thin white membrane. As the gall ages, the membranes break open to reveal a black powdery spore mass underneath. The spores are blown to adjoining corn plants where infection is repeated. Common smut of sweet corn is more of a problem during dry weather, which slows down the growth of the corn plant. Plants grown in soils high in nitrogen or plants damaged through cultivation are most susceptible to infection.

To control common smut, use resistant hybrids and plant high quality seed. Sweet corn should be grown on soil that is fertilised according to a current soil test recommendation.

Corn Stunt (spiroplasma): Corn stunt occurs in a small percentage in most sweet corn fields. It seldom reaches levels high enough to cause economic loss. The corn stunt spiroplasma is transmitted by leafhoppers. Infected plants are stunted, young leaves are yellow in color, and with age they take on a reddish-purple color. Internodes are reduced in length and infected stalks are sterile. Control is not required due to the very low percentage of plants that normally show this symptom in the field.

Maize Dwarf Mosaic (virus): Maize Dwarf Mosaic virus is the most common virus disease of sweet corn in Texas. Infected plants have mottled upper leaves that are lighter in color than healthy leaves. The mottled or mosaic pattern consists of alternate yellow and green islands in the leaf tissue. Aphids transmit virus particles from surrounding Johnsongrass. Johnsongrass rhizomes serve as the overwintering host for this virus. Early infected plants may be sterile. Late infection will reduce yields and quality of corn produced. Insect control is not successful due to the feeding pattern of the aphid. Elimination of Johnsongrass and isolation of sweet corn fields from Johnsongrass stands will help reduce the occurrence of this disease. There are a number of hybrids of sweet corn that are resistant.

EGGPLANT

Solanum melongena

Leaf Spot and Fruit Rot (fungus - Phomopsis vexans): This disease is characterised by circular brownish spots on fruit and leaves. On the fruit, soft, sunken spots become rotted and shrivelled. Spray with approved fungicide beginning when fruit is first set and repeat at 10-day to two week intervals until fruit is nearly mature. Use a three year crop rotation. Florida High Bush, Florida market and Florida Beauty are resistant. Early Blight (fungus - *Alternaria solani*): This disease can be destructive on eggplant at any time in the life of the plant. It can cause seedling dieback known as collar rot. Later infection is on the foliage beginning on the lower part of the plant and developing upward. Spots are characterised by concentric rings which give a target appearance.

Plants which are well fertilised and irrigated are not as susceptible. Infection of the fruit pedicels may cause a premature fruit drop. It requires eight to 10 days after infection before visible symptoms develop to the extent that epidemic levels are reached. Injured fruit are more subject to attack by the fungus than healthy fruit. Infection occurs between 60 and 90 degrees F. Long rotations, weed control, adequate fertiliser, and irrigation (furrow) will help reduce losses. Use clean seed and follow a thorough spray program when this disease is a problem.

Colletotrichum Fruit Rot (fungus - Colletotrichum melongenae): Lesions on the fruit vary from small spots to one-half inch in diameter. The tissue is sunken, with an area filled with a flesh-colored ooze of fungal spores. Spots vary from one to several on the fruit surface. Severely infected fruit drop to the ground with the pedicel still attached to the plant. The fungus overwinters in plant residue and grows at temperatures of 55 to 95 degrees F with optimum growth at 80 degrees F. Rainfall and overhead irrigation favor disease development. The fungus develops when the humidity is 93 percent or above. Although field sanitation is important, a preventive fungicide spray program is required during periods favorable for disease development.

Wilt (fungus - Verticillium albo-atrum): The pathogen attacks nearly 200 species of plants but eggplant and okra are the two most seriously affected vegetables. Young plants appear normal, but become stunted as they develop. Severely affected plants turn yellow. The lower foliage wilts and defoliation occurs. Symptoms continue to progress until death occurs. When the stem is cut, there is a dark brown, discolored band around the vascular system. Infection occurs directly through the root hairs. The fungus survives for indefinite periods in the soil. Survival is aided by weeds which are susceptible to the fungus. Infection takes place when the temperature ranges from 55 to 86 degrees F. Verticillium is favored in its development if the soil is alkaline. Some development takes place at pH of 5.0 but all growth is stopped at a pH of 4.0. Control involves the use of long rotations. Cotton gin trash should be avoided or be well composted if used.

Yellows (Tobacco Ring Spot Virus): The disease causes yellowing and whitening of upper leaves. Later, entire plant becomes yellow and may die. Avoid planting in fields where yellows have occurred and, if warranted, fumigate soil to control nematodes. The dagger nematode is a known vector of the virus.

ENGLISH PEA

Pisum sativum

Powdery Mildew (fungus - Erysiphe polygoni): Plants infected with this fungus are covered with a white powdery mold on the leaves, stems and pods. Infected plants are stunted and eventually die. The fungus is seed-borne. Optimum temperatures for development are between 68 and 75 degrees F. Soils should be kept as dry as possible yet will permit maximum growth. Avoid heavy application of fertiliser and rotate with non-related crops. Powdery mildew is the most serious disease of English peas in Texas.

Bacterial Blight (bacterium - Pseudomonas syringae pv. pisi): The bac- terium attacks all parts of the plant. Infected stems become olive brown in color, while the leaflets become yellowish or water-soaked. Young infected pods drop prematurely. Well formed pods become water-soaked.

The bacterium is seed-borne. Rain or irrigation water is necessary for movement of the organism. After infection takes place, four to six days is required before lesions are visible. Optimum temperature for development is 82 degrees F.

Control of this pathogen is through a combination of practices such as drainage, row spacing, seeding rates, weed control, and restricting irrigation only to that needed for maximum plant growth.

Ascochyta Blight (fungi - Ascochyta pisi, A. pinodella, A. pinodes): All three fungi are known to attack peas. Affected leaves have spots which are large, pale brown to dark brown in color. Lesions are papery in appearance and have gray to tan centers marked by small black pycnidia. Infection by some species cause purple lesions. On pods, deep lesions are formed which may have purple margins. The center is tan with black pycnidia.

Planting infected seed results in poor stands. All three species of fungi are seedborne and carry over in crop residue. Rainfall and heavy dews are necessary for infection. No infection occurs when the relative humidity is below 80 percent. Optimum temperatures for the fungus are between 68 and 82 degrees F.

Control is achieved by using clean seed, long rotations (four years or longer), planting in well-drained soil, and deep plowing to remove old crop residue.

Aphanomyces Root-Rot (fungus - Aphanomyces euteiches): Early infection causes complete crop loss due to seedling death. Late infection results in poor plant growth and reduced seed formation. Tissue decay does not develop above the soil line unless the weather is extremely wet. Infection occurs in both wet and dry soils, but is most destructive in wet soils. Optimum temperatures for infection are between 65 and 75 degrees F. The use of high levels of fertiliser will encourage continued root development. Nitrogen acts as a suppressant to fungal growth.

The use of three year rotation, well-drained soil, and the liberal use of fertiliser will help reduce losses from this disease. Mosaic (virus): Irregular, light and dark, greenish areas and puckering occurs in leaves. Control aphids that transmit the virus.

LETTUCE

Lactuca sativa

Downy Mildew (fungus - Bremia lactucae): First symptoms are yellowish or light green areas on the upper surface of the leaves. As the lesions enlarge, the fungus produces downy growth on the underside of the leaf. Affected tissue turns brown, sometimes comprising extensive areas of the leaf. Infected leaves are usually invaded by secondary bacteria and fungi causing considerable damage during shipping. The fungus is rarely seed-borne and lives in the soil for very short times. Infection takes place from diseased crop residue and is occasionally found on wild weed hosts. Spores of the downy mildew fungus are wind-borne and are carried considerable distances.

Spores germinate and infect leaves, producing symptoms within 5 to 10 days. Climatic conditions play an important role in the development of the disease. Cool, humid weather resulting from rain, heavy dew or fog is ideal for disease development. There are several races of the fungus and varieties differ in their reaction. Resistant varieties may become susceptible when new races of the fungus develop. Varieties resistant to known races should be planted. Preventative fungicide applications at frequent intervals will give satisfactory control of the disease.

Sclerotinia Drop (fungus - Sclerotinia sclerotiorum): The causal organism is a soilborne fungus that infects the stem of lettuce at or near ground level. Lower leaves wilt and collapse, the entire plant wilts and dies. Later, a soft, watery, brown rot destroys the plant. White, cottony growth of the fungus may be found on the surface of leaves and stem. Hard, black resting bodies called sclerotia, can be found in the cottony growth. The fungus is soilborne and can remain in the field for several years. The fungus first infects dead leaves at the base of the plant and later moves to live tissue. Under moist conditions, infection occurs when plants are reaching maturity. Avoiding infested fields is recommended.

Mosaic (virus): Young plants first show a slight leaf rolling, followed by light green to yellow mottling. Foliage has a general unhealthy yellowish color and plants are small or dwarfed. Symptoms vary according to type or variety of lettuce. The virus is seed-borne and overwinters in wild hosts and weeds. Infected seed is the most common source of infection. Once infected plants are present in the field, aphids further spread the virus. The virus can also be mechanically transmitted. There is no cure for mosaic once a plant is infected. Disease-free seed offer an effective means of control.

Big Vein (virus): Infection may appear at any stage of growth. Very young seedlings are killed. Surviving plants are smaller in size, quality of heads is affected and maturity is delayed. Infected leaves have an upright habit of growth and the veins turn yellow and begin to clear. Areas around the veins turn white causing the veins to appear enlarged.

The virus is soilborne and may persist in the soil for a long period of time. The virus can be transmitted by *Olpidium brassicae,* a soilborne fungus found in the rootlets of many growing plants. Avoiding excessive soil moisture and rotation are means of reducing disease losses.

Tip Burn (physiological): Margins of leaves die and turn brown. The first symptoms produced are small, dark brown spots about a quarter of an inch from the edge of the leaf. Tissue around these spots dies and turns brown along the margins. Tip burn usually becomes infected with secondary bacteria causing soft rot. The disease is believed to be related to fluctuations in soil moisture, drastic changes in temperature, and available calcium. No satisfactory control for this disease has been developed.

Damping-Off (fungi - Rhizoctonia sp., Pythium sp.): Small plants wilt and die soon after emergence. Excessive wet and prolonged cool, wet periods of weather are conducive to damping-off. Use only high quality seed which have been treated with a fungicide. Plantings should be on raised beds.

Soft Rot (bacteria - Erwinia carotovora): The bacterium enters through mechanical wounds or following damage caused by other diseases, insects or frost. Once the disease is established, it spreads rapidly in the field if warm weather and other conditions are favorable. There is no practical control of the disease once it becomes established and losses will increase during transit and marketing. Infected or damaged lettuce should not be packed; lettuce waiting to be shipped should be kept refrigerated.

Market Disease (various causes): Several disorders can occur on lettuce after harvest and during transport, which could easily be confused with disease caused by pathogenic organisms. Lettuce can be injured by excessive concentration of carbon dioxide (CO_2), low oxygen levels, exposure to ethylene gas, and fluctuations in temperature. Brown stains on leaves, spotting of midribs, rib discolorations and russet spotting are some of the symptoms usually associated with these disorders.

OKRA

Hibiscus esculentus

Fusarium Wilt (fungus - Fusarium oxysporum F. sp. vasinfectum): The most obvious symptom is a typical wilt, followed by death. Infected plants may be stunted and yellow. The fungus invades the root system and colonises the vascular system. In doing so, water movement is blocked and toxins from the fungus alter normal cell function. Cutting the base of the stem reveals a dark woody portion. No control is available other than a long rotation. All varieties are susceptible.

Root Knot Nematode (nematode - Meloidogyne sp.): Okra is highly susceptble. Root becomes enlarged and distorted.

Leaf Spot (fungi - Alternaria sp., Ascochyta sp., Cercospors malayensis, Phyllosticta hibiscina): There are several leaf spotting organisms which attack okra. However, none have been shown to cause economic loss. No control is recommended.

Blossom and Fruit Blight (fungus - Choanephora cucurbitarum): Young fruit and blossoms are attacked by the fungus which gives them a "whiskery" appearance. Infected plant parts are reduced to a soft rotten substance. Disease development is favored by warm, humid weather. Spray with an approved fungicide.

Seedling Disease (fungus - Rhizoctonia sp.): This disease is more likely to occur if okra is planted before soils warm sufficiently in the spring.

Virus: A whitefly-transmitted geminivirus was detected by Dr. Judith K. Brown in okra from Tamaulipas, Mexico in the fall of 1994. Fruit has irregular yellow areas which follow a longitudinal alignment. The initial symptom on young leaves is a diffuse, mottled appearance. Older leaves have irregular yellow areas which are interveinal. The disease is associated with the presence of whiteflies. The nature of the geminivirus has not been determined, nor has the presence of other viruses been ruled out.

Cotton Root Rot (fungus - Phymatorichum omnivorum): Infected plants die suddenly from mid-summer to fall. Leaves usually remain attached to the plant.

ONION

Allium cepa

Tip Blight (several causes): Several causes are known to result in the death of the leaf tips of onion plants. Some of these causes can be classified as physiological, while others could be the result of infection by soilborne diseases. Among the most common causes of tip dieback is overcrowding; when onions are planted thick, tip dieback is always present. Other causes of tip blight are damage of the root system by the disease known as pink root, damage by certain insects, particularly thrips and leaf miners, soil-related problems that place the plants under undue stress, and occasionally damage by ozone gas released into the atmosphere during severe thunderstorms accompanied by frequent lightning.

Botrytis Leaf Blight (Blast) (fungi - Botrytis allii, B. squamosa and B. cinerea): Botrytis leaf blight or blast is a serious disease of onion in other onion producing states, such as New York and New Jersey, where it appears almost every season. In Texas, its presence is more sporadic, although during the last few years it has been occurring with alarming frequency. Several species of Botrytis occur on onion leaves; the disease caused by *B. allii* is usually called neck rot; that caused by *B. cinerea* is known as leaf fleck, and the one caused by *B. squamosa* is known as leaf blight. The disease is frequently associated with small whitish spots occurring along the entire length of the leaf.

Most spots have greenish halos that at times appear to be water soaked. When the spots are numerous, the tip of the leaf may die back, giving the field a blasted appearance. However, other causes for these spots are suspected, such as cold rain, sleet, sandblasting and other fungal diseases. Most fungicides used for controlling purple blotch will also control Botrytis leaf blight.

Purple Blotch (fungus - Alternaria porri): The fungus affects leaves, seed, stems and bulbs. At first, small whitish, sunken lesions with purple centers are found on the leaves. These spots later enlarge and eventually encircle the leaf. Later, darkened zones appear on the surface of the leaves, retaining the characteristic purple color. Bulbs can be infected during harvest, curing, storage and transit. The disease will spread rapidly to other bulbs causing extensive damage. Recent research findings indicated that disease occurrence can be predicted accurately by measuring the number of hours that free moisture is present on the leaf surface. When 10 to 12 continuous leaf-wetness hours occur, the disease will develop when it is present. Fungicide application should be made when these conditions occur.

Downy Mildew (fungi - Peronospora destructor): Downy mildew is a relative new disease of onion in the Lower Rio Grande Valley, having made sporadic appearances during the last five years. When it does occur, it can cause severe economic damage to the crop. Symptoms consist of white to violet spots on leaves, later turning dark or almost black. Typical symptom of this disease as it occurs in Texas is a series of contiguous verticle spots up and down the leaves. A fuzzy growth is observed on the surface of the leaves, particularly during periods of high humidity. Fields should be monitored closely, particularly during prolonged cold wet weather, when the disease is more likely to occur. Fungicide applications should begin on plants two to three months old as soon as these conditions prevail. Downy mildew is seldom found in Texas as the season progresses into warmer weather.

Pink Root (fungus - Pyrenochaeta terrestris): Pink root is strictly a soilborne disease. Diseased roots turn pink in color, shrivel and die. As the plant sends out new roots, they become infected and die. Affected plants do not usually die, but bulbs are small and of poor quality. The disease affects only roots. Plants infected with the pink root fungus will also develop the tip blight condition discussed previously. The fungus usually is introduced to new land by transplants grown in infected soil. Once the soil becomes infested, the fungus remains in the soil for many years. Soil fumigation has been shown to be effective for the control of this disease. The cost, however, is high. The use of onion varieties resistant to pink root is the best control for this disease. Rotating fields will also reduce losses. Onions infected with the pink root fungus have also shown a higher incidence of *Fusarium* basal rot. Crop rotation is very important in attempting to control both of these problems.

Stemphylium Blight (fungus - Stemphylium vesicarium): The fungus causing this disease normally invades dead and dying onion tissue, particularly the tip blight and purple blotch lesions. This disease causes only minimal damage when plants are otherwise healthy. Control can be accomplished with the same chemicals used for purple blotch and tip blight control.

Neck Rot (fungus - Botrytis allii, Botrytis sp.): This disease is frequently unnoticed in the field because damage usually occurs during transit and storage. Diseased tissue at the base of the crown becomes sunken and watersoaked in appearance. A gray fungal growth later forms on the surface, which can be followed by other fungi and bacteria, causing decay. Careful handling of the crop at harvest and prompt drying of onions with heat and air ventilation are the best means of controlling this disease.

Basal Rot (fungus - Fusarium sp.): Disease occurrence starts at the basal plant of the onion bulb, causing the bulb to become soft. When onions are cut, a semi-watery decay is found, advancing from the base of the scales upward. The disease may remain unnoticed until transit, where the decay may continue until the entire bulb is destroyed. The fungus lives in the soil, invading the bulb through wounds or through root scars at the base. High soil temperatures favor disease development.

Black Mold Rot (fungus - Aspergillus niger): Black mold is generally destructive during storage and transit, although it may be observed on maturing onions in the field. The disease can be recognised by the presence of black powdery spore masses of the fungus on the outer scales. Observations indicate that high temperatures and moisture favor disease development. Bulbs should be protected from moisture in the field during and after digging and during transit.

Bacterial Soft Rot: Bacterial soft rot isone of the more prevalent causes of loss in storage onions. The soft rot bacterium can enter the neck tissues as plants approach maturity. As the rot progresses, invaded scales become soft and foul-smelling. Onions with mechanical injuries, sunscald, or bruises are particularly susceptible to bacterial soft rot, especially if they have been held under warm, humid conditions.

PARSLEY

Petroselinum crispum

Leaf Spots (fungi - Cercospora sp., Septoria sp., Alternaria sp., *Phylosticta sp.):* Small to medium size spots develop on leaves. They are usually yellowish when young and later turn brown. No fungicides are cleared on parsley.

SPINACH

Spinacia oleracea

White Rust (fungus - Albugo occidentalis): White rust is the major disease of spinach in Texas. White, blister-like pustules form on the lower side of the leaf. In advanced stages, the white lesions form on the upper side of the leaf. Generally, the upper surface will only be chlorotic. Plants infected with the white rust fungus are weak and collapse quickly if environmental conditions are favorable for disease development. Initial outbreaks each season follow source event such as hard rains that deposit soil and oospores on the young plants. The fungus oversummers in soil as dormant thick-walled oospores and may spread within a field by windblown spores. Free moisture on the leaf surface must be present for spore germination and development.

The optimum temperature for germination is 54 degrees F. The disease develops most rapidly at 72 degrees F. or during periods of cool, humid nights and mild day temperatures. No variety is totally resistant to white rust but a few have very useful partial resistance, including Green Valley II, Ozarka II, Coho and Fall Green. A disease control program should combine the benefits of cultural practices, partial resistance, a systemic fungicide at planting (in Texas, Ridomil 5G) and protective fungicides sprays (beginning 40-50 days after planting). Cultural practices should include long rotations, planting on beds and furrow irrigation. The risk of developing tolerance to Ridomil fungicide is reduced if all of these disease control practices are used together.

Downy Mildew or Blue Mold (fungus - Peronospora effusa): The downy mildew fungus first causes yellowish areas on the upper side of the leaf. The underneath side of the leaf is marked by a gray to violet-gray fungal growth mat that bears sporangia. The entire leaf is killed on susceptible varieties under optimum environmental conditions. The fungus overwinters in living spinach plants and in the seed. The fungus spores require surface moisture for development. Optimum temperature is around 48 degrees

F for germination and 54 to 60 degrees F for development. This disease can be controlled by the use of varieties resistant to races 1, 2 and 3.

Anthracnose (fungus - Colletotrichum spinaciae): Anthracnose shows up as small, dark olive colored spots. As the spots enlarge, they become tan in color. As the lesions coalesce, they kill the entire leaf. During periods of favorable disease development, the foliage appears as if the crop will be lost. With a change in weather, the diseased foliage drops off and the healthy foliage goes on to produce a crop. The fungus overwinters in seed and in crop refuse. Fungicides used in a white rust control will also control anthracnose.

Leaf Spots (fungi - Cercospora beticola, Heterosporium variable): These are minor diseases of spinach which may cause damage in some years. *Cercospora* leaf spots are white and, usually, small in size. Extended periods of precipitation and high humidity may allow spots to become large and even coalesce. Protective type fungicides offer some control. Ridomil does not control *Cercospora.*

Heterosporium leaf spots are larger and have a greenish black fungal growth on both sides of the leaf as the disease develops. Control measures are not generally recommended.

Fusarium Decline (fungus - Fusarium oxysporum f. sp. spinaciae): Plants are subject to infection at any stage of development. Infected plants turn yellow and wilt. The lower, older foliage is affected first. Plants which are infected are either stunted or killed. Infected plants lose their feeder roots and the vascular system of the taproot is darkened. The fungus is seed borne and can live in the soil indefinitely. Fortunately, white rust resistant varieties also have *Fusarium* resistance.

Blight (virus): Cucumber mosaic virus attacks spinach causing plants to become faintly chlorotic. Chlorosis increases until the entire plant becomes yellow. Crown leaves are narrow, wrinkled and develop an inward rolling of the margins. Plants are stunted. Death may occur in some severely infected plants. The virus is aphid transmitted. Most commercial varieties are resistant.

Seedling Disease (fungi - Fusarium spp., Rhizoctonia spp., Pythium spp., Others): Preemergence and postemergence damping off can seriously reduce stands. Rotate with corn, plant after soils have cooled down in the late summer or fall, use a fungicide seed treatment and buy fresh, high quality seed.

Tobacco Ringspot (virus): First symptoms are small, indistinct, chlorotic spots which appear on the young foliage. These may coalesce to form large yellow areas. In advanced stages, leaves take on a copper bronze chlorosis. There is no malformation of the foliage. Affected plants rarely die but remain yellow and stunted. No control measures are known.

Beet Curly-Top (virus): Infected plants are marked by a rosette of tightly curled, small leaves in the center of the plant. As the disease develops, the growing point is killed and the plant dies. The virus is transmitted by beet leafhoppers.

False Root Knot (nematode - Naccobbus crucifera): Plants are stunted; in cases of severe early infection, death will occur. The root system of infected plants is characterised by large galls covered with numerous rootlets. This nematode is particularly damaging on spinach plants in the fall. This nematode does little damage in hot soils.

Stress (abiotic - heat and/or saturated soils): The quality of spinach can decrease quickly following stress. Yellowing occurs within a few days after high temperatures and/or flooding. Choose planting dates carefully for your areas and provide good drainage.

SWEET POTATO

Ipomoea batatas

Scurf (fungus - Monilochaetes infuscans): Scurf is also known as soil stain. It is most severe in wet or poorly drained soils. Scurf appears as light brown areas on the outside of the roots and are only skin deep. These areas may be small or run together to form large, irregular patches. Although superficial, infection reduces grade and causes undue shrinkage in storage. Avoid use of scurfy seed potatoes. Treat sweet potato roots with fungicide before planting. Practice rotations of three to five years. Avoid planting where organic matter in the soil is not well decomposed.

Root Knot (nematode - Meloidogyne sp.): Root cracking is often associated with severe root knot infection. Pitting and other surface blemishes are also noted. The cracking is sometimes confused with growth cracks. The splitting of small whip-like roots with small black surface lesions will reveal tiny, pearly-like nematodes embedded around the blackened areas. These also occur in mature potato flesh but are more easily viewed for diagnostic purposes in smaller roots. Vines may be stunted and show nutritional deficiencies. Root knot can be controlled by selecting only healthy seed roots, free of root knot nematodes. These should be planted in a bedding site that is also free of the nematode and preferably has not been used in many years. Preplant nematicides or soil fumigants will help control field infestations. Most newer varieties such as Jewel, Cordner and Topaz have some degree of resistance.

Stem Rot (fungus - Fusarium oxysporum f. sp. batatas): Young leaves at the tips of vines turn yellow while older leaves wilt then drop so that the center of the hill becomes bare. Infested slips may die soon after setting or become stunted and yellow. Stems at the soil line may turn slightly blue (blue-stem). The inner stem portion at or below the soil line becomes discolored with brown streaks in the vascular system. Use of stem cuttings rather than slips or sprouts for planting stock will reduce the likelihood of new infections in the field. Control by using resistant or tolerant varieties. Selection of seed roots free of the disease is difficult because external symptoms are not always present. Limited control may be obtained when treating seed with a fungicide for other soilborne diseases.

Black Rot (fungus - Ceratocystis fimbriata): Circular, almost black spots appear on the sweet potato. Affected and adjacent areas have a bitter taste. Small, black lesions often completely girdle underground stems. The causal organism fruits abundantly in storage and spores are easily spread by mites so that a few diseased roots may result in severe storage losses. Select seed potatoes carefully at harvest time and again at bedding. Use new soil each season for bedding seed. Cull all slips that appear sickly and those with black spots on lower stems. Crop rotations, clean plant beds and use of a fungicide on planting seed will also aid in controlling black rot. If black rot occurs in storage, the rotted roots should be destroyed and the storage area disinfected.

Soil Rot or Pox (fungus - Streptomyces ipomoea): Potatoes are often mis- shapen with rough, scabby pits or shallow surface lesions that result in scars. Soil rot is most severe in soils approaching alkalinity. Control may be obtained by soil fumigation. Some varieties have tolerance to soil rot.

Foot Rot (fungus - Plenodomus destruens): Foot rot is not as widespread as black rot or stem rot and, therefore, losses to this disease are not as great. On infected vines, the base of the stem turns brown at the soil line and leaves nearest the crown turn yellow and drop. It is most commonly observed from mid-season to harvest. Individual plants may produce few potatoes even though large vines develop during the season. The potatoes that are produced develop a firm brown rot at the stem end. Seldom is the entire root affected. Controls recommended for black rot and scurf will also control this disease. If foot rot is recognised to be a serious problem, then early harvest will aid in reducing losses.

Internal Cork (virus): Foliar symptoms vary from purple ring spots to vein-clearing. Symptoms become masked as leaves grow older. The most characteristic symptom is seen within potatoes since no visible surface symptoms are evident. Slicing a potato lengthwise will reveal brownish to black corky areas inside. If seed potatoes are stored at 60°F., development of the corky areas does not proceed. However, if seed are stored at 70 degrees F., severe corkiness develops. Storing seed potatoes at 70 degrees F. is one way of checking for this disease. The variety Puerto Rico is highly susceptible. Internal cork should not be confused with chilling damage.

Storage Rots (fungi - soft rot - Rhizopus stolonifer, Java black rot - Botryodiplodia theobromae, black rot - Ceratocystis fimbriata, charcoal rot - Macrophomina phaseolina and other): Storage rot losses are greatly reduced when disease control practices are followed that yield high quality sweet potatoes from the field. Some fungi causing storage rots infect roots in the field before harvest while others enter the potatoes through wounds made at harvest or during handling. Decayed spots may be dry or soft, the latter due principally to Rhizopus rot.

Sweet potatoes should be cured for seven to 10 days at 85 degrees F. and at 90 percent relative humidity before being stored. Curing allows a natural healing process to take place. Potatoes should be stored at 50 to 60 degrees F. and at 85 to 90 percent relative humidity. Avoid rough handling between curing and storage because additional wounding may occur. Store in a thoroughly clean and disinfected room with adequate ventilation. Do not permit the storage temperature to drop below 55 degrees F. or injury will occur. For control of Rhizopus rot during the processing-to-market interval, apply a fungicide spray immediately after washing, when potatoes are on conveyor belts or rollers.

Cold Damage (physiological): Pithy, dark discolorations will appear internally on sweet potatoes subjected to low temperatures. Temperatures do not have to reach the freezing point to cause damage. Sweet potato is a tropical plant and will suffer injury at temperatures below 55 degrees F.

(physiological): Scientists have not yet determined the conditions that cause cracking of sweet potatoes. In some varieties, it appears to be an inherited characteristic. Rotation reduces incidence of growth cracks.

TOMATO

Lycopersicon esculentum

Late Blight (fungus - Phytophthora infestans): Lesions produced on the leaves are at first irregular, rather large, greenish-black and water-soaked. These areas enlarge rapidly, becoming brown, and under humid conditions, develop a white moldy growth near the margins of the diseased area on the lower surface of the leaves or on stems. The disease spreads rapidly under humid conditions, destroying quickly large areas of tissue. Fruit lesions occur as large, green to dark brown lesions, mostly on the upper half of the fruit, but they may also occur on other parts. White moldy growth may also appear on fruits under humid conditions.

The fungus produces abundant numbers of spores which may be splashed by rains or be airborne. These spores infect healthy leaves, stems and fruit readily if climatic conditions are optimum. Ideal conditions for late blight development are cool nights and moderately warm days with abundant moisture. Hot and dry weather reduces disease development. Infection on tomatoes may occur from nearby potato fields or from airborne spores. Control practices include rotating fields so as not to follow potato or tomato; avoiding planting tomatoes near potatoes; using disease-free seed and transplants. Crops should be sprayed regularly with a fungicide. Spraying should begin whenever weather conditions are favorable for disease development. Applications should continue on a 7 to 10 day schedule until harvest.

Early Blight (fungus - Alternaria solani): Early blight is first observed on the plants as small, black lesions mostly on the older foliage. Spots enlarge, and by the time they are one-fourth inch in diameter or larger, concentric rings in a bull's eye pattern can be seen in the center of the diseased area. Tissue surrounding the spots may turn yellow. If high temperature and humidity occur at this time, much of the foliage is killed. Lesions on the stems are similar to those on leaves, sometimes girdling the plant if they occur near the soil line. Transplants showing infection by the late blight fungus often die when set in the field. The fungus also infects the fruit, generally through the calyx or stem attachment.

Lesions attain considerable size, usually involving nearly the entire fruit; concentric rings are also present on the fruit. The fungus is cosmopolitan, affecting a wide range of plants and is present in and on the seed and on crop refuse. Infection of young seedlings usually results from fungi and soil; it occurs principally during periods of high humidity and warm temperature. Control practices include the use of disease-free transplants and regular preventative fungicide applications during the growing season.

Gray Leaf Spot (fungus - Stemphylium solani): First infection appears as minute, brownish-black specks on the lower leaves that extend through to the undersurface of the leaf. These spots usually remain small, but may enlarge until they are about one-eighth inch in diameter. They become glazed and at times the centers crack and tear across. Infected leaves usually die and drop. Spots also form on the stems. The fungus overwinters primarily on diseased refuse; spread is by airborne spores from diseased to healthy plants. Control measures include crop rotation, seedbed sanitation and preventative fungicide applications as for the control of late and early blight. Resistant varieties should be used when possible.

Leaf-Mold (fungus - Fulvia (Cladosporium) fulvum): Leaf mold is usually first observed on older leaves near the soil where air movement is poor and humidity is high. At first, diffuse whitish spots appear on the upper surfaces of older leaves; these rapidly enlarge and become yellow. Under humid conditions, the lower surface of these spots become covered with a gray, velvety growth of the spores produced by the fungus. When conditions are proper for fungal development, large areas of the field are infected, plants are weakened and the crop is greatly reduced. The fungus produces abundant spores during periods of high temperature and very high relative humidity. Infection occurs readily, and the disease becomes established in the fields quickly. The best control of this disease is by using a preventative fungicide program at 7 to 10 day intervals, the same as used for late and early blight control.

Buckeye Rot (fungus - Phytophthora parasitica): This disease occurs on tomato mainly on the fruit, particularly where it touches the soil. The fungus is different from the one causing late blight, which affects both leaves and fruit. Buckeye rot is first noticed as a light green water-soaked area on the fruit. Later, dark zonate bands can be seen on the surface of affected areas. The surface of the lesion is usually smooth and firm. With time, the entire fruit will rot. The fungus lives in the soil and it can also affect pepper. The disease is more troublesome in heavy, poorly drained soils during prolonged warm wet weather.

Bacterial Spot (bacterium - Xanthomonas campestris pv. vesicatoria): Infected leaves show small, brown, watersoaked, circular spots about one-eighth inch in diameter. The spots may have a yellow halo; the centers dry out and frequently tear. On older plants the leaflet infection is mostly on older leaves and may cause serious defoliation. The most striking symptoms are on the green fruit. Small, water-soaked spots first appear which later become raised and enlarge until they are one-eighth to one-fourth inch in diameter. Centers of these lesions become irregular, light brown and slightly sunken with a rough, scabby surface. Ripe fruits are not susceptible to the disease. Surface of the seed becomes contaminated with the bacteria, remaining on the seed surface for some time.

The organism may also overwinter in alternate hosts, on volunteer tomato plants and on infected plant debris. Moist weather and splattering rains are conducive to disease development. Most outbreaks of the disease can be traced back to heavy rainstorms that occur in the area. Infection of leaves occurs through natural openings; infection of other fruits must occur through insect punctures, sandblasting and other mechanical injury means. Bacterial spot is difficult to control once it appears in the field. Disease-free seed and seedlings should always be used; rotate crops so as to avoid last year's crop residue; use a combination of copper and organic fungicides in a regular preventative spray program at 5 to 10 day intervals. Dry, hot weather usually checks the disease.

Bacterial Canker (bacterium Corynebacterium michiganense): Wilting of leaflets is the first symptom; on older plants the margins of the lower leaflets wilt first, the margins dry and the leaflets curl upward. Light colored streaks appear on stems. These streaks later break open at intervals to form cankers in which bacteria ooze may appear. Fruits are infected on the surface, spots appearing first as very small and snowy-white; the margins remain white and flat, but the centers become raised, tan colored and roughened. A white halo develops around the spots, which serves to differentiate the disease from other bacterial spots. The canker organism is seedborne and can remain in the soil and diseased plants for some time. Primary infection, however, is mostly from infected seed. Control can be obtained by using disease-free treated seed and clean transplants.

Nailhead Spot (fungus - Alternaria tomato): Leaf symptoms are the same as those caused by early blight on fruits; however, spots are smaller, with slightly sunken centers and dark margins. As the spots become older, the edges become roughened. On ripe fruit, the tissue immediately around the spots often remains green. Control is the same as for early blight.

Anthracnose (fungi - Colletotrichum sp.): At first, infected fruit show small, slightly sunken, watersoaked spots. These spots enlarge, become darker in color, depressed and have concentric rings. Masses of the pink fruiting fungus can be seen on the surface of the lesions in moist weather. Under warm and humid conditions, the fungus penetrates the fruit, completely destroying it. The fungus persists on infected plant refuse in the soil. Fruit may be infected when green and small, but do not show any marked lesions until they begin to ripen. Fruit becomes more susceptible as they approach maturity. Control of this disease involves the use of well-drained soil, crop rotation and a preventative fungicide program as recommended for other diseases.

Fusarium Wilt (fungus - Fusarium oxysporum f. sp. lycopersici): The first indication of disease in small plants is a drooping and wilting of lower leaves with a loss of green color followed by wilting and death of the plant. Often leaves on only one side of the stem turn yellow at first; yellowed leaves gradually wilt and die. The stem of wilted plants shows no soft decay, but when cut lengthwise, the woody part next to the green outer cortex shows a dark brown discoloration of the water conducting vessels. The fungus is soilborne, passes upward into the xylem of the stem. Blocking of the water-conducting vessels is the main reason for wilting. The fungus is most active at temperatures between 80 and 90 degrees F., seldom being a serious problem where soil and air temperatures remain low during most of the growing season. Control can be obtained by growing plants in disease-free soil, using disease-free transplants, and growing only resistant varieties.

Verticillium Wilt (fungus - Verticillium albo-atrum): The first symptom is yellowing of the older leaves, followed by a slight wilting of the tips of the shoots during the day. Older yellowed leaves gradually wither and drop, and eventually the plant is defoliated. Verticillium wilt does not show the one-sided effect as does Fusarium wilt. Leaves from *Verticillium* infected plants sometimes show brown dead spots that may be confused with those caused by other fungi. However, they are lighter in color and do not show concentric zones as in early blight. In late stages of the disease, only the leaves near the tips of the branch remain alive. When the stem is cut lengthwise, the base shows a discoloration of the woody tissue similar to Fusarium, but is usually darker, and generally it occurs only in the lower part of the stem. The fungus enters the plant through the feeder roots and grows into the stem in the woody conducting vessels just under the cortex. The fungus lives in the soil for a long time and it is exclusively the source of infection. Progress of the disease is favored by cooler temperatures and is retarded by the high temperatures that are favorable to *Fusarium* wilt. Locating seedbeds and fields in *Verticillium*-free soil, and using resistant varieties are the most effective means of controlling the disease.

Gray Mold (fungus - Botrytis cinerea): Plants become more susceptible to this disease as they become older. It is mostly a problem in greenhouses, but it can also affect tomatoes in the field. The fungus first becomes established on dead leaves at the base of the plants. A heavy, gray growth of the fungus covers these, and numerous spores are soon found, giving the affected area a cottony appearance. Affected leaves collapse and shrink. The fungus progresses into the stem, producing cankers. Affected fruits first show a watersoaked, soft area in the points of infection. The dark gray growth of the fungus soon is seen on these spots. Regular fungicide applications should help in controlling this disease.

Botryosporium Mold (fungus - Botryosporium sp.): This fungus can often be found on greenhouse tomatoes. It superficially resembles gray mold. Septoria Leaf Spot (fungus - *Septoria lycopersici*): Infection usually occurs on the lower leaves near the ground, after plants begin to set fruit. At first, small watersoaked spots are observed, which under ideal conditions will become numerous. Large areas of the leaves may be affected but the individual spots can be recognised. The watersoaked spots become roughly circular, with dark margins surrounding a light gray center. With time, black specks which are spore producing bodies can be seen in the center of the spots. If the spots are numerous, the lower leaves will turn yellow, die and progressively drop from the plant until only a few leaves remain on the top of the plant. The fungus is most active when temperatures range from 60 to 80 degrees F., and humidity conditions are high. The disease is usually not serious during periods of hot, dry weather. The fungus can overwinter on crop residue from previous crops, decaying vegetation and some tomato-related wild hosts. Crop rotation, plowing under crop residues, and clean cultivation will reduce the amount of inoculum in tomato fields. Repeated fungicide applications will keep the disease in check.

Southern Blight (fungus - Sclerotium rolfsii): The first symptom is dropping of leaves suggestive of other wilts. Wilting progresses and plants die quickly. Stems show decay of outer tissues at the ground line. Frequently, they are covered by a white fungal mat in which are embedded numerous small, light-brown bodies about the size of cabbage seed. The fungus can also attack fruits where they touch the soil. The fungus does not grow at temperatures below 68 degrees F.; it requires abundant moisture for growth. Infection takes place below the soil line or close to ground level. Control is obtained by sanitation, crop rotation, and by treating infected soil with a soil fungicide prior to planting.

Blossom End Rot (nonparasitic): Blossom end rot is a physiological disorder of tomato that can appear on fruits at any time in their development, but most commonly appears when fruits are one-third to one-half grown. The initial symptoms are water-soaked spots on the blossom end of the fruit. These spots later enlarge and become black. Secondary infection by other decay causing organisms usually follows. The cause of this disorder is considered to be calcium deficiency in the developing fruit. Extreme fluctuations in moisture, root pruning and excessive nitrogen fertilisation can also result in blossom end rot. Best means of control is to maintain a uniform supply of moisture through irrigation and soil mulches. Calcium chloride can be used as a spray for control under some conditions.

Growth Cracks (nonparasitic): Growth cracks result from extremely rapid fruit growth brought on by periods of abundant rain and high temperatures, especially when these conditions take place following periods of stress. Cracks of varying depth radiate from the stem end of the fruit, blemishing the fruit and providing an entrance for decay-causing organisms. No control practices are known.

Tobacco Mosaic (virus): Symptoms are light and dark green mottling of the tomato foliage, and curling and slight malformation of the leaflets. Plants may be somewhat stunted if infected when small, but the plants and fruit are not much reduced in size if plants are not infected until they reach the fruiting stage. Several strains of the virus are known that can cause different symptoms. The virus is highly infectious and readily transmitted by any means that introduces even a minute amount of sap from infected to healthy plants. The most common means of transmission is by handling contaminated plants. The virus may also be present in certain types of tobacco; therefore, smokers may transmit the disease. Control measures are: avoiding handling plants more than necessary, washing hands before handling plants, and protecting healthy plants from infection.

Double Streak Virus (virus): Caused by a combination of viruses. Leaves show a light-green mottling, accompanied by the development of numerous small, grayish-brown, dead spots which have a thick, paper appearance. Numerous narrow, dark brown streaks develop on the stem and leaf petiole. Fruits are often rough and misshapen and on the surface they have small, irregular, greasy, brown patches which render them unfit to market. The virus is transmitted by workers handling the crop. Avoid infection by tobacco mosaic virus, wash hands before starting to work and remove infected plants.

Spotted Wilt (virus): This disease is similar to streak in that it causes streaking of the leaves, stems and fruits. Numerous small, dark, circular spots appear on younger leaves. Leaves may have a bronzed appearance and later turn dark brown and wither. Fruits show numerous spots about one-half inch in diameter with concentric, circular markings. On ripe fruit these markings are alternate bands of red and yellow. The virus also affects other vegetables and many wild hosts and ornamental plants. Thrips can transmit the disease from the wild hosts. For control, eliminate weeds around field edges and turnrows; remove infected plants when small, and control insects in the field.

Curly Top (virus): Pronounced upward rolling and twisting of the leaflets that expose their undersurfaces, stiff and leathery foliage, and a peculiar dull yellowing of the entire plant are typical symptoms of the disease. There is also some purpling of the veins and the plant is usually very stunted. Very few fruits are produced after infection. The virus is not transmitted through the seed or soil, nor is it spread by mechanical means. The main vector is the beet leafhopper that becomes infected by feeding on wild or cultivated plants having the disease. The disease is difficult to control. Keep field surroundings free of weeds. Controlling insects may effect some control.

Root Knot Nematodes (Meloidogyne spp.): Root knot nematodes represent one of the major limiting factors to tomato production. Preventing their entry into growing areas is an important mean of control. Susceptible varieties growing in infested soil will have knots or galls on the root system. Once present it becomes necessary to use a resistant variety or to apply nematicide. Refer to the section on nematodes for more information. Use of resistant varieties is the most effective method of reducing nematode losses.

Greenhouse diseases—All of the diseases affecting tomatoes in the field can damage tomatoes in the greenhouse. If proper precautions are not taken, severe losses can occur to tomatoes grown under greenhouse conditions. Among the most common diseases of tomatoes under greenhouse conditions are:

Soil diseases: Nematodes, Fusarium wilt.

Stem diseases: Bacterial cankers, Botrytis stem canker, early blight canker.

Foliage diseases: Leaf mold, gray mold, late blight, early blight, mosaic. It is important that special control practices are practiced under greenhouse conditions. Maintaining the proper relative humidity is essential for the control of foliage diseases. Relative humidity should be 85 percent or less, which at times proves to be very difficult. Proper air circulation, ventilation, and temperature control are essential under greenhouse conditions. Sanitation is extremely important in controlling greenhouse diseases. All debris should be collected and destroyed soon after harvest. Tools and equipment should be properly cleaned at periodic intervals. Hands should be washed with soap and water before entering the house, and smoking should never be allowed. The soil should be fumigated before planting to control nematodes and other soilborne organisms. Whether using gas or liquid fumigants, proper rates and methods of applications should be made periodically to control foliage diseases. Some fungicides can be sublimated by fire under airtight conditions. If properly used, they will effectively control these diseases.

TURNIP AND MUSTARD

Brassica spp.

White Spot (fungus - *Cercosporella brassicae*): Lesions caused by *Cercosporella* varies from gray to brown in color. The center of the spots are light colored with slightly darkened margins.(Symptom) Severely infected leaves turn yellow and drop. The fungus is seedborne and overwinters on volunteer plants. Spores are spread by splashing rain and airborne spores. The disease is most damaging when daily temperatures range between 55 and 65 degrees F and during periods of high humidity. Protective type fungicides should be applied on a weekly basis beginning when true leaves are formed if weather conditions are favorable for disease spread. Cultural practices that reduce leaf wetness and inoculum levels will aid in disease control.

White Rust (fungus - Albugo candida): The underside of infected leaves are characterised by white blister-like pustules. (Lower Leaf Symptom) The upper side will have faint yellow spot opposite the white lesion on the underside of the leaf. Cruciferous weeds such as wild mustard and shepherd's purse are other hosts for the fungus and are a source of inoculum for mustard and turnips growing nearby. The fungus is spread by windblown spores and overwinters in crop refuse. Disease development is most rapid at 68 degrees F and during periods of light rain or dew. Control of the disease is based on cultural practices. Currently no fungicides are approved by EPA for the control of this fungus on mustard or turnips.

Anthracnose (fungus - Colletorichum higginsianum): Infected leaves are covered with small dry, circular, pale-gray to straw colored lesions.(Leaf symptom) Severe infection results in leaf death. On infected roots gray to tan lesions are formed. The bacterial soft rot organism often enters the lesions as a secondary infection and causes additional loss. The fungus overwinters in fallen leaves, on volunteer plants and some weeds. It can be seed transmitted. Optimum fungus growth occurs in a temperature range from 79 and 86 degrees F. The presence of moisture is important in disease development. Long crop rotations, weed control, and a well drained soil are important to reducing losses to this fungus. Approved fungicides can also be used when the disease appears. Begin fungicide applications when symptoms first appear and continue at 7 to 10 day intervals as long as weather conditions favor disease development.

Alternaria Leaf Spot (fungus - Alternaria brassicae): Small circular spots with a yellow halo are formed on the upper leaf surface. (Leaf Symptom) As the spot matures the center becomes sooty black with age. The black color is partially the result of spore buildup on the surface of the lesion. Lesions are characterised by concentric rings within the lesion. Roots may also become infected when leaves drop on the soil. The fungus grows best at warm to hot temperatures and in the presence of light rainfall and dew. Spores are carried under the seed coat. Cultural practices will help reduce losses to this fungus disease. Apply fungicides at first sign of leaf symptoms.

Downy Mildew (fungus - Peronospora parasitica): The disease is most damaging in the fall and late spring, early summer months. The first symptoms are small faint yellow spots on the upper foliage. As they mature, fungus fruiting structures are visible on the underside of the leaf. This growth is especially noticeable in the morning following a period of mild temperatures, high humidity and no wind. Petiole and stem lesions are indistinct, elongated and gray to black in color. In rainy weather, bacterial soft rot may invade plants weakened by downy mildew. Fungicides can be used as preventative sprays. It is important to monitor fields closely for signs of disease build up. Follow cultural practices to reduce inoculum build up and spread.

Black Rot (bacterium - Xanthomonas campestris pv. campestris): Black rot infects a large number of cruciferous plants including several commercially important vegetables. Infected mustard leaves are marked by a "v" shaped lesion on the leaf margin. The bacteria enters through the water pores at the leaf margin and then move downward in the vascular strands of the leaf. In black rot infected leaves, the vascular system appears to have black to dark gray strands running through the tissue. Once infected leaves rapidly collapse. Infected leaves after harvest continue to break down in storage. The bacterium is carried in seed and can survive for a short time in the soil. Disease spread within a field is by rain, irrigation water or mechanical injury. Follow a least a 1 year. Longer rotations with non-cruciferous crops are suggested if the disease has been especially severe. Cultural practices can be used to reduce infection. Apply bactericides at first sign of leaf infection. Repeat at weekly intervals as long as weather conditions favor disease development. Damage will be first visible on lower foliage.

Bacterial Leaf Spot (bacterium - Xanthomonas campestris pv. raphani): The bacteria that causes bacterial leaf spot is seedborne. Foliage symptoms are pinpoint, watersoaked, olive green lesions. As they enlarge the lesions become translucent spots with watersoaked margins. The bacteria not only are spread on infected seed but can possibly survive in the soil and on crop residue for a short period of time. The bacteria are spread by splashing rain and equipment. Cultural practices that reduce the time foliage is wet help to reduce infection. Rotations with vegetables other than members of the crucifer family will help to reduce presence of the bacteria in the soil. Fungicides used for black rot control will also help to control this organism.

Turnip Mosaic (virus): Leaves of infected plants are mottled and distorted. The entire plant is stunted and yields are reduced. Aphids transmit the virus from diseased plants and weeds to healthy plants. Field sanitation is important. At the end of harvest, diseased plants or those plants not harvested should be shredded and plowed under or tilled to reduce inoculum increase in an area. If diseased plants are left in place, aphid will continue to increase and as feeding pressure increases, winged forms develop and spread the virus to nearby mustard or turnip plantings. Control cruciferous weeds in and around the field or garden. Volunteer mustard and turnip plants should be destroyed.

REFERENCES

Bailey, L. H., *Manual of cultivated plants,* The Macmillan Company, Now York.

Kezer, A., and W. G. Sackett, *Beans in Colorado and their diseases,* 1918.

Thompson, H. C., *Vegetable crops,* McGraw-Hill Book Company, Inc., New York. 1923.

5

Diseases of Fruit Crops

DISEASES OF APPLE

Apple Blotch (fungus - Phyllosticta solitaria): Shiny, black blotches develop on fruit. Edges of lesions are irregular with radiating outgrowths. Foliage symptoms are characterised by light gray spots with a dark dot in the center. Elongated, sunken, light colored lesions with black dots are formed on veins, midribs and petioles of the leaves. The spread of blotch is favored by temperatures of 75 to 80 degrees F and heavy rains. Several fungicides will control this disease

Bitter Rot (fungus - Glomerella cingulata): Symptoms appear first as small, circular, brown spots on both apples and pears. The lesion enlarges and becomes saucer-shaped. The fungus overwinters in mummified fruit and cracks or crevices in bark. Most infections can be traced to a broken limb. Infection normally shows up in mid to late summer. Young fruit are resistant to the fungus. Maximum infection occurs at 85°F and after light rain. Orchard sanitation and a regular spray program are extremely important. Broken limbs should be cut to prevent overwintering of spores in these protected areas.

Black Rot (fungus - Physalospora obtusa): Black rot is a firm, textured rot. The spot is light brown and darkens with age. As it enlarges, it is marked with concentric dark bands. The lesion is convex in shape, yet in very advanced stages becomes somewhat depressed. On the leaf the lesion is a purple speck that enlarges and eventually has a center which is brown or yellowish-brown. The fungus infects foliage at petal fall. Fruit infection is not normally evident until midsummer. The disease is favored by temperatures of 80 degrees F and light rain. Although fungicides help, sanitation is the best means of reducing this disease.

Bot Rot (fungus - Botryosphaeria ribis): The fungus infects woody tissues and fruits. Wounded apple wood is very susceptible. On limbs and twigs, infection is visible in July. It appears as a blister on the infected limbs. At maturity the blister ruptures and a liquid spreads over the limbs. The wet area then becomes sunken with a dark colored appearance. The spot spreads into the cambium where it enlarges in concentric rings until fall. Lesions stop enlarging in the fall and cracks appear along the edge of the spot. It may or may not begin growth next spring.

Fruit infection is observed as small reddish-brown spots. These eventually coalesce, causing a soft rot to occur. As the decay develops, the lesion turns light brown to dark brown and is covered with beads of exudate on the surface of the decayed fruit.

The fungus overwinters in live and dead limbs. In the spring, an ooze is formed which is spread over the tree. Mature and ripened fruit are very susceptible. The disease develops at temperatures of 75 degrees F or above. Botryosphaeria infection *is encourag*ed by any condition which reduces tree vigor such as drought, winter injury, and low nutrition. Jagged pruning wounds and grafting sites are sometimes invaded. Pruning out the decayed areas and a thorough spray program will significantly reduce losses to this disease.

Cedar Apple Rust (fungus - Gymnosporangium juniperi-virginianae): The fungus passes part of its life on apple and part on the cedar. On cedar it forms reddish galls which swell and produce an orange slime each spring. On apple it causes spots on leaves, fruit and sometimes on young twigs. The spots on the foliage and fruit are orange-yellow in color, and a blister or cushion forms in the center of each spot. Where this disease is a problem, cedars should be destroyed for 1 mile around an apple orchard. Normal spray programs will control infection on apples. Resistant varieties should be used when available. Golden Delicious and Delicious are resistant. Most ornamental crabapples and Jonathans are susceptible.

Cotton Root Rot (fungus - Phymatotrichum omnivorum): Apples should never be planted in soil infected with this fungus due to its extreme susceptibility.

Fire Blight (bacterium - Erwinia amylovora): This bacterium causes twig dieback and is common on apples and pears as well as some 75 other plant species. Blossom blight is common on apples. Infected blossoms wilt and turn black rapidly. The foliage does not immediately abscise due to the rapid rate of death. Twig blight is common and can result in dieback ranging up to 12 to 36 inches of twig length. The bacteria overwinters in cankers. Bacteria oozes from the cankers in early spring and are carried by wind and insects to healthy foliage. Resistant varieties should be used when available. Fungicides containing copper will cause russetting on apples. Trees which have had fire blight should be pruned to remove overwintering cankers. Nitrogen levels should be kept low. No organic nitrogen sources should be used as they encourage continual growth which is susceptible to the bacterium. Make cuts from 4 to 6 inches below the visible cankers. All pruning equipment should be sterilised with a one part household bleach to nine parts water solution.

Fly Speck (fungus - Leptothyrium pomi): Numerous small, shiny, black spots are formed on the fruit. Control is the same as sooty blotch.

Mushroom Root Rot (fungus - Armillaria mellea): Mu*shroom root rot* is caused by a soilborne fungus which attacks apples and causes a slow decline. Trees may be infected with the fungus for 2 to 3 years before death occurs. The vigor of the tree is noticeably reduced during this time. Trees infected with Armillaria mellea have *a layer of creamy* white fungal growth between the bark and the wood at the soil line. The growth (rhizomorphs) girdles the trunk and the tree dies when this girdling is complete. Trees damaged by the fungus can be diagnosed by the presence of the rhizomorphs and the advanced stage of decay of the root system. Infected trees are loose in the soil due to the deterioration of the root system.

Mushroom root rot is most often found in areas where post oaks were the predominate native timber and in sandy soils. This does not always hold true, and growers must be aware that the disease can cause serious losses in many soil types should it become established.

Soil fumigation will give only temporary control of this problem. The wide distribution of the fungus in the soil (both vertically and horizontally) and the constant recontamination of a fumigated area from nonfumigated soils reduce the early advantage gained by soil fumigation. Resistant root stocks are not available. Growers should avoid planting in areas where the fungus has been a problem in the past or on recently cleared post oak timberland.

It is best for a homeowner to move to another planting site should plants die from mushroom root rot.

*Phytophthora Collar Rot (fungus - Phytophthora cactorum): Black, slimy canker*s are formed from the bud union to the lower portion of the trunk. Trees infected with this fungus are stunted and eventually die. It is particularly damaging on the Malling Merton Clones. Most seedlings of Delicious, Melba, McIntosh, and Wheatly are resistant. No chemical control recommended.

Powdery Mildew (fungus - Podosphaera leucotricha): Young foliage shows cupping and twisting. The leaves are covered with a white, powdery mass of spores. Net russetting will occur on fruit when it is infected with this fungus. All varieties are susceptible, yet Jonathan appears to be more susceptible than most. Use of a regular fungicide program will control this problem. Karathane, a commonly recommended fungicide, should not be used at temperatures of 90 degrees F or above. A spreader should be used to improve control.

Quince Rust (fungus - Gymnosporangium clavipes): This fungus attacks both apple and pear. It attacks the blossom end of the fruit and restricts the growth of the apple. Tissue below the lesion dies and becomes spongy. Quince rust, like cedar apple rust, must complete part of its life cycle on cedar. Apple varieties susceptible to quince rust include Delicious types. Use fungicides and remove cedars in the area for control.

Scab (fungus - Venturia inaequalis): This fungus attacks both the foliage and the fruit. It first appears on the leaves as small, dull, smoky areas which with age become olive colored, velvety, and much more visible. On fruit, the lesions are first small, circular, olive green areas which turn dark and scabby. The spots crack open in late stages of development.

The fungus overwinters in fallen apple leaves. The fungus produces fruiting structures in the spring. Rains cause spores to be shot into the air where they are carried to developing foliage. The spore will germinate and produce a leaf spot if moisture is present when the spore lands on the leaf. Further spread occurs from these spots to adjoining leaves of fruit. Apple scab is dependent on moisture for development.

Sooty Blotch (fungus - Gloeodes pomigena): Sooty-gray or cloudy blotches appear on the fruit. The fungus attacks both apples and pears in Texas. It is most severe during years that have cool, wet springs, late summer rains, and low temperatures in early fall. It is normally found in conjunction with fly speck, another disease of apples and pears. Fungcides should be applied in late spring and early summer.

Southern Blight (fungus - Sclerotium rolfsii): Infected trees die rapidly (2 to 3 weeks) after the first visible symptoms are observed. The leaves remain attached to the tree. Creamy white to yellow rhizomorphs are formed on the outside of the roots. As the growth develops, small, dark brown sclerotia are formed among the rhizomorphs. In Texas it has been observed attacking apple trees both in the nursery and in commercial orchards.

Currently, no chemicals are cleared for the control of Southern Blight on apples. Deep burial of organic material prior to planting and controlling weeds at the base of the tree will help reduce losses. Sclerotium rolfsii is *a facultative sapr*ophyte which grows on organic material in the soil, but under certain conditions, it can attack and parasitise healthy growing plants. Crop residue should be pulled away from the base of the apple trees.

Avocado Diseases

Anthracnose (fungus - Colletotrichum gloeosporioides): Black circular spots up to one-half inch appear on fruit. The center of the spots may be slightly sunken and spots may develop cracks. During moist periods, the spots produce pinkish, moist masses of fungal spores. As the fruit ripens, the infection spreads rapidly into the flesh causing a greenish-black, fairly firm decay. Fungicides can control the disease and should be applied at bud swell.

Cotton Root Rot (fungus - Phymatotrichopsis omnivora): Although avocado is susceptible to cotton rot root, it is not a common disease. It is characterised by a sudden wilt of the tree. The leaves turn brown and remain on the tree. Very young trees tend to become diseased, rather than old ones. The only control is to avoid planting trees in soil known to be infested with the fungus.

Leaf Burn (physiological): Leaves brown at the tips and the edges. Affected leaves may drop prematurely. These symptoms can be caused by accumulation of salts in the soil, inadequate soil moisture, wind dessication, and frost. Light irrigation should be avoided since this fails to leach accumulated salts out of the root zone. The soil should have good drainage.

Phytophthora Root Rot (fungus - Phytophthora cinnamomi): Infected trees have small leaves which are lighter green in color than healthy leaves. The leaves wilt and drop, eventually leading to complete defoliation of the tree. Twigs and branches die back. There is a light fruit set with small fruit. The fungus infects roots up to one-quarter inch in size, although it is the feeder roots that are primarily affected. These roots become blackened and brittle before they die.

The disease occurs if the pathogen is present in soil and if there is excessive soil moisture. Poor drainage can be a contributing factor to disease development. The fungus can be introduced to new areas by movement of infected nursery stock or infested soil. Soil on implements can also serve as a means for introduction of the fungus to non-infested areas. Fungicides should be used in combination with sanitation in the nursery and maintaining adequate drainage in the field.

Scab (fungus - Sphaceloma perseae): Circular, brown, scabby areas are found on mature fruit. Leaves have brown spots and become crinkled. The disease can become a problem if there is cool, moist weather when fruit and leaf tissue is young. Fungicides can control the disease and should be applied when flower buds appear, near the end of the main bloom period, and 3-4 weeks after that.

Seedling Blight (fungi - Phytophthora spp.): This can be an important disease in nursery production. Leaves show irregular reddish-brown areas that enlarge along the larger veins. The terminal bud may be killed. The disease is favored by periods of heavy rainfall and high humidity, when seedlings are young and succulent. The following cultural practices can prevent or minimise damage. Plants should be grown on benches. The use of soil should be avoided to prevent introduction of the pathogen. Irrigation water from ponds may also serve as a pathogen source. Providing good air circulation between plants will provide a less-favorable microclimate for disease development. Fungicides should be used only in combination with cultural control practices.

Sunblotch (avocado sunblotch viroid): Twigs have a light yellow, sunken streak that follow the lenght of the twig. Fruits have white or yellow areas. Trees are stunted and have a sprawling growth habit. The pathogen is transmitted in budwood, graftwood, or seed from infected trees. Disease-free nursery stock should be used and trees with symptoms should be removed from the orchard.

Wood Rot (fungus - Ganoderma lucidum): This fungus contributed to the gradual decline and death of trees in a portion of an orchard in the Rio Grande Valley. The decline occurred over a period of several months during the summer. The leaves of affected trees first turned pale to yellow and trees soon lost leaves. Eventually, trees were completely defoliated and died. Fruiting bodies of the fungus were found on dead and dying trees. The above-ground symptoms of the disease resemble Phytophthora root rot. The disease occurred in a portion of the orchard with poor drainage that was subject to waterlogging. The waterlogging led to the death of feeder roots (again, a symptom that resembles Phytophthora root rot) and stressed the trees. This environmental stress allowed the fungus to develop within trees. The fungus infected the trees via wind-blown spores that entered the tree through wounds. There is no chemical control for this pathogen. Encouraging vigorous tree growth by minimising environmental stresses (e.g. improving soil drainage) will prevent establishment of the fungus in the tree.

DISEASES OF BLUEBERRY

Botrytis Blight (fungus - Botrytis cinerea): Under high humidity and mild temperatures this fungus can attack blooms and tender growth. Blossoms may shrivel prematurely as if injured by frost. The fungus grows profusely, producing a gray to brownish fuzzy material on infected parts. Rarely is Botrytis a serious problem, but when it occurs, heavy losses can result. Botrytis can also cause post harvest fruit rot.

Leaf spots (fungus - Gloeocercospora inconspicua, Septoria albopunctata): Leaf spot diseases develop on occasion but are considered minor problems in production plantings. Fungicide treatments are not generally recommended.

Mummy Berry (fungus - Monilinia vaccinii - corymbosi): The fungus infects when growth begins in spring, causing a leaf and flower blight. It might be confused with Botrytis blight at this stage. Blighted leaves and blossoms quickly develop a light brown fungal growth on external surfaces as they dieback. Some infected blossoms continue development without noticeable symptoms. As these fruit approach maturity, they turn a pink to white color instead of blue. By this time the fungus has colonised the fruit, forming hard, black, survival structures inside that remain dormant until the following spring. Mummy berry disease is not common, but when it occurs, there can be heavy losses, especially during bloom. Preventive fungicide treatments are recommended for control.

Stem Blight (fungus - Botryosphaeria dothidea): Leaves yellow and infected branches die. Rarely is the entire plant lost. Prune out diseased wood. Fungicides have not proven beneficial.

Stem Canker (fungus - Botryosphaeria corticis): Although susceptible, most rabbiteye varieties have some degree of resistance to stem canker. Like stem blight, stem canker eventually causes branch dieback. Prune out affected wood. Fungicides are not effective.

Other diseases: Other diseases that may occur on blueberry are Phytophthora root rot, powdery mildew, crown gall and ripe rot (fruit rot near maturity).

DISEASES OF COMMON SWEET CHERRY

Bacterial Canker (bacterium - Psuedomonas syrinigae pv. syringae): This bacterium causes cankers, gummosis, and shoot blight. Sunken lesions appear in the spring with gummosis. Cracks and cankering follow, resulting in a girdled stem. The foliage becomes yellow, curled, and withered. Wet spring weather can result in shoot blight and a small angular leafspot. This is usually only a problem in young orchards. Use resistant F12-1 Mazzard rootstock. Remove girdled branches. Spray with copper hydroxide in October and January. This bacterium also attacks peaches.

Black Knot (fungus - Apiosporina morbosum): Black knot is a very conspicuous and common disease that is also found on apricots and plums. Large, rough black galls form on the twigs. Twigs may die beyond the galls. No chemical control is recommended. Infected limbs should be removed.

Brown Rot (fungus - Monilinia fructicola and Monilitia laxa): These fungi cause blossom blight, twig blight, and fruit rot.

Crown Gall (bacterium - Agrobacterium tumefaciens): Warty tissue masses or galls may form on the roots, at the root collar, or anywhere on the stem.

Leafspot (fungus - Wflsonia hiemalis): This is the most common and most damaging leaf disease of cherries. It is most prevalent during rainy springs. The reddish spots drop out leaving circular holes in the leaves. With severe infection, defoliation may follow the appearance of the "shot hole" symptom. This disease can be controlled with foliar fungicides.

Little Cherry (virus): Fruit are half size. Flowering cherry acts as a reservoir for the virus. Buy only virus-free trees from reputable nurserymen and remove infected trees when they are discovered.

Littleleaf: This is a condition in the West caused by a lack of zinc.

Rusty Mottle (virus): Foliage and blossoms are delayed in the spring. Brown necrotic or rusty chlorotic spots develop on the leaves, often falling out to leave shot holes. Defoliation may occur. Buy virus-free trees from reputable nurserymen. Virus infected trees should be removed from the orchard.

Rugose Mosaic (virus): General chlorosis of the leaf occurs between mid vein and margin with distortion. Fruit yield is reduced. The fruits become flattened and angular. Remove infected trees.

Rust (fungus - Tranzschelia discolor): Rust also occurs on peach and plum. Yellow spots appear on the upper leaf surface with reddish pustules filled with spores on the lower surface. It can cause premature leaf drop lowering tree vigor. Apply fungicides in late summer and fall.

Rusty Mottle (virus): Many leaves tum bright yellow to red with green islands and drop before harvest. Those that remain have yellow-brown spots with a rusty appearance. The fruit is small, late, and tasteless. Remove diseased trees and select grafting material from virus-free trees.

Shot-hole or Black Leaf spot (bacterium - Xanthomonas pruni): This bacterium also attacks peach and plum. The infected tissue dries up and falls out leaving a one-eighth inch diameter hole. It also causes a canker and gummosis of the stem.

Vein Clearing or Sweet Cherry Crinkle (probably genetic): A non-transmissible virus-like disease. Veins clear in localised areas or the entire leaf Margins of the leaves are irregular. Some are elongated with slot-like perforations. Small blisters appear on lower sides of veins, and the upper sides are silvery. Leaves fold along midrib, wilt, and drop in midsummer. Some branches will rosette. Flowers are plentiful, but few will develop into fruit. The fruit will be pointed and flattened on one side with a swollen ridge.

Witches' Broom (fungus - Taphrina cerasi): This fungus causes large numbers of irregular small branches to develop on twigs and larger branches forming a witches' broom. Blossoms develop and the leaves come out on the brooms earlier than on the normal branches. Large numbers of brooms may eventually kill a tree. Cut and burn brooms.

DISEASES OF CITRUS

Black Rot (fungus - Alternaria citri)

The infection starts while the fruit is still on the tree. The fruit is infected through cracks or openings on the stylar end of fruit. The fruit colors prematurely and drops.

Brown Rot (fungus - Phytophthora spp.)

The same species of Phytophthora that causes foot rot can infect fruit, commonly those close to the ground. Fruit are infected during periods of excessive rain, resulting in decayed areas that are brown, firm and leathery. Later, a white, velvety growth is seen on the surface of the fruit, accompanied by a strong fermenting odor. The best control for brown rot is sanitation during transit and in the packing house. Disinfecting the boxes is important. Treatment with disinfectant solutions and refrigeration are effective for prevention.

Cachexia (Xyloporosis) (viroid)

This pathogen makes the inner bark surface bumpy. The bark projections are smooth, in contrast to the sharp projections produced by the citrus tristeza virus. Cachexia affects certain mandarins and tangelos, particularly Orlando tangelo. Sour orange is tolerant to the virus.

Citrus Tatter Leaf (tatter leaf-citrange stunt virus)

This virus causes a bud union crease. It can be detected using certain indicator plants (e.g. swingle citromelo citrus). It is serious on rootstocks of Troyer and Carrizo.

Cotton Root Rot (fungus - Phymatotrichopsis omnivora)

This is a minor disease of citrus and it rarely appears on trees that are more than three years old. The fungus kills trees quickly; most of the dried leaves remain attached to the branches.

Diplodia Stem-End Rot (fungus - Diplodia natalensis, also known as Physalospora rhodina)

Fruit are contaminated with the fungus in the field, but the disease commonly occurs at the packing house or in transit. Decay occurs around the stem end and advances in streaks down the side of the fruit. There is no fungal growth on the surface of the fruit. Decay is reduced by dipping the fruit in fungicide solutions before placement in de-greening rooms. Promptness in handling and shipping, as well as refrigeration during transit, will minimize losses.

Exocortis (citrus exocortis viroid)

This pathogen causes bark-shelling and stunting of trees on trifoliate orange, trifoliate hybrids, and Rangpur lime rootstocks. Sour orange rootstock is tolerant. In the early stages of the disease, gum exudes from pustules at the base of the trunk and may extend from below the soil line to the bud union. New bark forms beneath the pustules, and the outer bark sloughs off forming the characteristic bark-shelling. The tree eventually declines. The pathogen is spread via grafts and also via pruning tools. The pathogen can be detected following grafting on Etrog citron. Symptoms on this indicator plant include stunting and leaf deformation and necrosis of the leaf midrib.

Feeder Root Rot (fungi - Phytophthora spp.)

The fungus infects the cortex of feeder roots, giving the root system a stringy appearance. This can lead to yield loss and a general prolonged tree decline. Sour orange rootstock is relatively resistant. This disease can be controlled with fungicides. Soil solarisation may also be an effective treatment.

Flyspeck (fungus - Leptothyrium pomi)

This is caused by a fungus that keeps citrus fruit from turning yellow in the infected spots. Small black specks are formed on the rind in areas immediately surrounding the oil glands. The contrast between green and yellow color lowers fruit grade, even though there is no effect on fruit or juice quality.

Firing (physiological - wind dessication)

The injury is usually limited to late growth flush (September or October) which has not matured prior to the arrival of a dry cool front in October or November.

Green Mold (fungus - Penicillium digitatum)

The fungus causes a rapid breakdown of fruit punctured or bruised during harvesting and packing operations. The fungus enters the fruit only through wounds. A white mold that later turns green is seen growing on the skin. The decayed fruit becomes soft and shrinks in size. Losses from green mold can be reduced by preventing injury during harvest and transit. The fruit is treated in the packing shed with a fungicide solution.

Greasy Spot (fungus - Mycosphaerella citri)

This is a very common disease in Texas. The fungus is primarily a leaf pathogen, but it can also infect fruit. Yellow spots are seen at first on the upper surface of the leaf. The corresponding areas on the undersides of the leaves turn dark and appear slightly raised and greasy. With severe infection, leaves may turn yellow and drop prematurely. The infection is favored by rust mite injury and humid weather. Yellowing caused by scale insects may be mistaken for early symptoms of greasy spot.

Either citrus oils or fungicides can be applied during the summer for control. Heavy infections occurring in wet years may require 2 to 3 fungicide treatments. Since infection occurs on the lower leaf surface, thorough spray coverage is essential for effective control. The spores of the fungus originate in the leaf litter on the orchard floor and are released following rainfall or irrigation. Thus, removing litter will reduce the amount of spores present.

Heart Rot/ Wood Rot (various fungi)

The visible external symptom is the fruiting structures ("conks") of the fungus projecting from the trunk. Eventually, the tree declines. Species documented on citrus in Texas include Ganoderma lucidum, which has a shiny, brown appearance on the upper surface of the conk, and Phellinus robustus, which appears flat on the bark surface.

These fungi, and other species, cause decay of internal wood. The fungi enter the tree through injuries to the roots or scaffold limbs or improper pruning cuts. Growth of the pathogen is favored when the tree is under environmental stress. Once infected, the tree can not be treated. Proper pruning practices, which promote rapid wound healing, will minimise the risk of infection.

Leaf Miner (insect)

The larvae tunnel within the leaf, leaving snaking trails of dead tissue.

Leaf Yellowing (physiological - various causes)

Injury by the herbicides Krovar (a mixture of diuron and bromacil) or Hyvar (bromacil) can cause yellowing. Magnesium deficiency causes yellowing of the leaf, except for a lanceolate area along the petiole. However, this type of yellowing can also be caused by iron or zinc deficiency, or it may be a consequence of foot or root rot.

Melanose (fungus - Diaporthe citri)

This is primarily a disease of grapefruit. The first symptoms on leaves are small, circular, dark depressions with a yellow margin. Later, the spots become raised and turn dark brown. Leaves turn yellow and may drop prematurely. Raised spots are also found on twigs, which may be killed. Fruit can also become infected. The disease can be severe following rainy periods in the spring, particularly when such periods follow a freeze that has left an abundance of dead twigs.

Melanose is usually controlled by a single fungicide application after bloom or petal fall and before the fruit averages one-half inch in diameter. With a particularly wet spring, a second application may be needed. Citrus should be sprayed for melanose control in the spring following a freeze. Chemical control efforts in heavily infected orchards may need to be combined with supplemental pruning to reduce the quantity of the pathogen.

This fungus also infects fruits. Spots on the fruit are at first small, light brown, and sunken; later they become dark and raised. When several spots are close together, the surface has a rough feeling to the touch, hence the name "sandpaper" melanose. Spots sometimes develop in a tear-streaked pattern, resulting from infection caused by spores which wash down over the fruit surface during heavy dews or light rains. In other cases, large areas of the fruit surface crack in irregular patterns, resulting in the "mud cake" type of melanose.

Oleocellosis (physiological)

The damage is caused to rind cells by peel oil release as a result of abrasion, rough handling or other injuries. Foggy and wet weather predispose rind to oleocellosis by making the fruit more turgid. The first symptoms are spots, which later become sunken. The damage can be minimised by harvesting later in the day on days with cool, foggy mornings.

Phytophthora Foot Rot/Gummosis (fungus - Phytophthora spp.)

This is one of the more important pathogens of citrus in Texas. It primarily infects the roots and the trunk. Infection of the trunk results in a dark, water-soaked areas, often with profuse exudation of a dark resin from the lesion. The dead bark frequently sloughs off the wood in vertical strips. If the lesion encircles the trunk, girdling occurs, leading to the death of the tree. Trees killed by this fungus are defoliated and show an abundance of fruit. Trees with the bud union beneath or close to the soil and trees in poorly drained locations are especially susceptible to infection.

Phytophthora foot rot can be prevented through the use of resistant sour orange rootstock, budded 4 to 6 inches high, and planting at the proper depth in well-drained soil. Flood irrigation water must be kept below the scion and micro-irrigation emitters should be placed so that the trunk is not left wet for prolonged periods. Soil banks should be removed as soon as the danger of frost has passed. When tree wraps are used on young trees, trunks should be treated with a recommended fungicide and they should be inspected for bark damage or gumming if there has been fire ant activity.

Pruning and sprout removal are best done only in dry weather. Cultural activities that result in occasional trunk injury, such as disking or boom spray applications should be done at least 2 days in advance of irrigation.

Trunk lesions can be treated with fungicides. Severely damaged or unproductive trees should be removed from the orchard.

Psorosis (Scaly Bark) (virus)

Sweet orange, grapefruit, and tangerine are severely affected. The pathogen is spread by budding nursery stock with budwood from infected trees. Variable symptoms can be produced on young leaves and can include pale flecks or mottling. The earliest bark symptom is scaling. Bark scaling of the trunk and larger branches is typical of psorosis symptoms in citrus trees eight to 12 years or older. As the disease progresses, the tree declines rapidly and eventually becomes unproductive.

Rio Grande Gummosis (fungus - Physalospora rhodina, also known as Diplodia natalensis, possibly other fungi involved)

This is not a common disease. Narrow cracks appear in the bark which exude a pale yellow gum. Symptoms are usually found in association with injured or dead branches. The wood beneath infected tissue is pink to orange in color. Later, gum pockets develop beneath the bark. Stressful growing conditions, such as freeze damage, high water table, salt accumulation, and poor cultural practices predispose trees to the disease. Removal of dead wood and pruning which promotes quick healing is recommended as a preventative measure.

Rust mite (insect)

The fruit can be scarred as a result of mite feeding. Early infestation results in a sharkskinned fruit texture of tan color, and fruit size is reduced. Later infestation causes russetting or bronzing of the rind.

Slow Decline (nematode - Tylenchulus semipenetrans)

The nematode causing this disease attacks the roots. Symptoms associated with the presence of nematodes are wilting, lack of vigor, poor fruit production, small fruit size, and poor response to watering and fertilisation. Poor drainage, excess salt, improper watering and damage caused by other soilborne organisms will affect trees in much the same way as the citrus nematode.

Positive diagnosis requires laboratory analysis of root and soil samples, which are obtained following specific sampling guidelines. Losses caused by the citrus nematode in heavily infected orchards can be reduced by applying nematicides to the soil. The table of citrus nematode action thresholds can be used as a guide to determine whether nematode populations warrant chemical controls.

When establishing a new orchard, avoid old sites known to be infested. Soil should be sampled before establishing a new orchard to determine if parasitic nematodes are present in damaging numbers. Ideally, nematode-free plants should be used when establishing new orchards, i.e. grown in containers or in fumigated nursery soil.

Sooty Mold (fungi - Capnodium spp.)

Leaves, fruit and sometimes branches have a black, moldy appearance. The fungi causing sooty mold do not actually infect the plant, instead they grow on the sugary exudates of insects such as aphids, brown soft scale, blackflies and whiteflies. The amount of sooty mold found is directly proportional to the number of honeydew-secreting insects present.

A heavy sooty mold coating on the fruit can result in a lower grade of fruit, while a heavy coating of fungi on the leaves could retard growth, causing lighter flowering and reduced yields. Insect control will prevent sooty mold development. Heavy sooty mold can be treated by adding one-half percent oil to the insecticide spray.

Sunscald (physiological - strong sunlight)

Irregular, brown pustules are found on the underside of leaves. This occurs when the undersides of leaves are suddenly exposed to direct sunlight, for example, after harvest because of repositioning of limbs and leaves after removal of the fruit that had bent them downward.

Tristeza (citrus tristeza virus)

The virus infects most types of citrus and it is transmitted by certain aphids or by the use infected budwood. The symptoms produced depend upon the variety, environmental conditions, and the virus strain. The virus can cause a decline or death of the tree. Mild strains were identified in Texas in 1991, although tristeza was also detected in the 1950's. There are several serious strains that are not known to occur in Texas. Three strains are of particular concern: one causes the decline of sweet orange on sour orange by killing phloem, resulting in girdling at the bud union; and the stem pitting strains (one on grapefruit, the other on sweet orange). The stem pitting strains causes limbs of trees to become brittle, forms deep pits on the underside of bark, and a roping appearance of the trunk that results in a decline of the tree. Although the tree doesn't die, there is reduced set, size and quality of fruit.

The virus is inefficiently vectored by three species of aphids present in Texas: Aphis gossypii, A. spiraeola, and Toxoptera aurantii. It is very efficiently vectored by the brown citrus aphid, Toxoptera citricida, which is present in Florida, but not in Texas. If the brown citrus aphid arrives in Texas, it could move severe strains of tristeza suspected to be present on symptomless hosts in dooryard (i.e. non-commercial) trees to commercial orchards, potentially destroying the Texas industry. This is because sour orange rootstock, which is predominantly used in Texas, is susceptible to severe strains.

Control efforts in Texas are mainly directed toward keeping infected budwood out of the Lower Rio Grande Valley by quarantine measures and a shoot tip grafting and budwood certification program to clean up infected material already in Texas and place it in commercial orchards. Should these measures fail, a proposed alternative control is cross-protection of trees with mild strains of the virus and/or alternative rootstocks which may tolerate the disease.

Twig Dieback

Twig dieback can be caused by several fungi, as well as by other factors. Fungal infection often occurs following a freeze or mechanical or chemical injury. Affected young branches die back from the tip, sometimes producing gum exudation. Wood is discolored underneath the bark. Damage by twig dieback usually is not severe.

Wind Scar

This damage occurs as the result of twigs or leaves rubbing against young fruit. The injury develops early, i.e. within the first couple of months post-bloom.

DISEASES OF FIG

Anthracnose (fungus - Glomerella cingulata): The fungus which causes anthracnose attacks both the fruit and the foliage. Infected fruit are characterised by a soft rot and premature dropping of the fruit. Immature fruit are dried up and may remain on the tree. Infection results in a small, sunken, discolored area. The areas enlarge with age and become covered with a pink mass of spores. Affected leaves will have a dark brown margin. Defoliation occurs with increased infection. Sanitation is extremely important in the fig planting. Diseased fruit as well as infected leaves should be removed.

Dieback (physiological - cold injury): Fig trees are often injured by early or late frosts that kill younger twigs. Although their death is not related directly to loss in production, they may serve as a site for secondary fungi to get started. All dead twigs and limbs should be pruned from the trees.

Fig Mosaic (virus): Affected figs show large yellow areas in the leaves, oak leaf pattern, ring spot area, or a mild mottled pattern. Leaves may be smaller than normal and deformed. Premature defoliation and fruit drop often occur. The virus is spread by vegetative cuttings and *Aceria ficus* (eriophyid mite). Control is by selection of clean propagating stock and insect control.

Fig Rust (fungus - Physopella fici): The disease is first evident as small, angular, yellow-green flecks on the leaf. The spots do not become extremely large but do become more yellow and finally a yellowish-brown. The margin of the spot is reddish in color. On the upper surface the spots are smooth, while on the lower surface the spots appear as small blisters. Brown spores are released from the blisters at maturity. As infection continues, the leaves become more yellow, and finally they begin to die around the leaf margins. Eventually death and defoliation occur. Complete defoliation can occur in two or three weeks. Fig rust generally becomes a problem as the fruit reaches maturity. Therefore, fungicide applications should be started in the early spring when the first leaves are completely grown. Make additional applications as new growth is formed. Do not spray when the fruit is one-fourth inch in diameter as the spray residue will make the fruit unattractive. Resume spraying after the fruit has been harvested.

Fruit Drop (physiological - flower development): The fig produces four types of flowers (male, female, Gall, and Mule): The male and female flowers are most often associated with the Capri type fig. This fig requires a wasp for pollination. The wasp does not occur in this part of the United States, thus it is impossible to grow Capri figs in Texas. Gall flowers are imperfect female flowers. They are found only on Capri and Cordelia figs. Mule flowers need no pollination and produce no seeds. The common fig grown in Texas produces primarily mule flowers. Since no seed are formed, the mule flowers are more subject to dropping than than those flowers which require pollination. The presence of the seed and the growth hormones produced by the seed help prevent fruit drop. The figs grown in Texas due to absence of seed are more subject to premature fruit drop as a result of adverse growing conditions.

Leaf Blight (fungus - Pellicularia kolerga): In early stages of infection, small areas in the leaves become yellow and appear watersoaked. With continual development, the upper surface becomes silvery white, and the lower surface becomes light brown and covered with a thin fungal web. In most cases, the leaves will turn brown and shrivel. It affects primarily the leaves but may develop on some fruit if it is new and a severely affected leaf or stem tip. Sanitation is the only recommendation to reduce losses from this disease.

Limb Blight (fungus - Corticum salmonicolor): Affected limbs wilt rapidly. The fungus enters at a spot along the main or secondary limbs, and all leaves die beyond that point. The fungus enters at a dead fruiting spore or at some other injured spot. All dead twigs and limbs should be removed by pruning so that they will not serve as infection sites.

Root Knot Nematodes (nematode - Meloidogynespp.): Root knot is one of the most common disease problems occurring on figs. Infected roots are characterised by small galls or swellings on the roots. The presence of the galls on the roots interferes with the normal uptake of nutrients by the roots. Plants infested with root knot are stunted and have a general unhealthy appearance. Infested planting sites should be treated with Vapam prior to planting. This will reduce the nematodes in the soil to a low level. Do not use around living plants as it will result in severe root pruning, and in many cases death will occur. Make sure the fig plant is free of root knot. Once planted, the only practice left is to keep the plant in good health with regular fertiliser applications and maintain adequate moisture around the plant. If nematodes were initially present, the fig will eventually become infested, but the root system should be well established by then.

Sclerotium Blight (fungus - Sclerotium rolfsii): A yellowish-white mat of fungal growth is formed at the base of the plant. Round, hard, yellowish to brown bodies (sclerotia) are found scattered in the fungal growth. To prevent the occurrence of this disease, it is important to carry out a thorough sanitation program. Old leaves or grass around the base of tree will encourage fungal development.

Souring (several fungi and bacteria): Organisms are carried into the fruit by the dried fruit beetle. Figs which have open "eyes" or ostioles should not be planted. Only those with closed "eyes" should be planted. Some examples of closed eye figs are Celest, Texas Everbearing, and Alma. No chemical control has been found to be totally effective.

Maneb fungicide will help to some extent. Insects should be controlled to eliminate them as carriers for the disease causing organisms.

Diseases of grape Anthracnose (fungus - Elsinoe ampelina): This disease is often called "birds-eye-rot" because of the circular, sunken, ashy gray, dark-bordered spots on the berries. It attacks fruit stems, leaf veins, petioles, tendrils, and young shoots in addition to the berries. The berry often cracks to the extent of exposing the seed. The disease overwinters in old lesions. Periods of high humidity favor disease development. Fungicides used for black rot will help prevent this problem.

Black Rot (fungus - Guignardia bidwellii): Black rot shows up as small, reddish-brown spots on the upper leaf surface. In older lesions the margin is a black line while the inner area of the spot is brown. Small, black dots are also visible in the center of the lesion. Infected fruit shrink until they are dried mummies. The first stage of development on the fruit is small, light-colored lesions with black borders. In advanced stages the fruit is marked with the small, black dots just like the foliage. Young grape foliage is most susceptible to this disease. Disease development is favored by high temperatures and humidity. A rainy period followed by two to three days of foggy weather favors development of this disease. Preventative fungicides along with resistant varieties should be used to control black rot.

Chlorosis: Chlorosis in grapes is usually caused by iron deficiency. American varieties are particularly prone to have this problem. The leaves turn yellow, but the veins remain green. If not remedied, chlorosis will decrease the yield, reduce sugar content of the fruit, and eventually kill the vine. Two applications of iron sulfate or iron chelate during the growing season should control chlorosis. In high pH soils, iron sulfate and some iron chelate may become tied up and unavailable to the plant. Foliar applications can be made, or a chelate especially for alkaline and calcareous soils, such as Sequestrene 138Fe, can be used.

Crown Gall (bacterium - Agrobacterium tumefaciens): Crown gall is caused by soil-inhibiting bacteria. The bacterium often enters wounds caused by freeze injury or hail. It stimulates the cells in the plant tissues to grow rapidly, resulting in the formation of tumor-like, spongy overgrowths which develop around the crown of the plant. Rain can splash bacteria onto the aerial parts of the vine where the tumors are often called black knots. The pathogen can move in the vessels and spread throughout the vine. Sanitation is the best prevention. Avoid cutting into affected tissue when pruning. Chemical and biological controls are now available.

Dead Arm (fungus - Eutypa armeniacae): This fungus invades grape tissue slowly. Spring symptoms on developing shoots and leaves adjacent to an infected pruning stub may not appear for four to five years. Cankers form on vines, and nearby leaves become chlorotic and dwarfed. Cordons die later, producing the "dead arm" symptom.

Downy Mildew (fungus - Plasmopara viticola): Downy mildew is common in cool, humid environments but seldom occurs in hot, dry areas. Downy mildew is first observed as a pale yellow area on the upper surface of the leaf. The underside of the leaf is marked by a downy appearance. As the disease advances the infected tissue dies and turns brown. Young stems become thickened and are often covered with the white fruiting structures. Fruit that is infected is covered with the white growth or it turns the berry a dull green and then brown. Downy mildew is a particular problem in areas of high humidity. The disease develops in temperatures of 50 to 60 degrees F. Some American grape varieties show resistance, but additional chemical protection is usually needed. Most European varieties are very susceptible. Fungicide application should begin before bloom and continue at seven day intervals.

Grapevine Fanleaf (virus): Fanleaf is caused by a virus that also causes yellow mosaic and veinbanding. All three are transmitted by the nematode *Xiphinema index*. Leaf symptoms that resemble a fan are very conspicuous on Mission and French Colombard. Infected vines have shortened and more irregular internodes. Lateral sprout development, double nodes, and stem faciations cause a bushy appearance. Many berries shatter, and others do not develop beyond shot size. Diseased vines should be removed.

Leaf Roll (virus): Leaves roll downward and turn red between the veins progressing toward the cane tips. In California, symptoms appear in early June in non-irrigated vineyards and in August in irrigated vineyards. Vines have fewer, smaller clusters per vine with berries that are low in sugar. Red fruit varieties, such as cardinal and Mission, develop fruit lacking color, and berries of white grapes, such as Thompson seedless and Reisling, develop a yellowish-white color instead of the normal greenish-white. In general, leaf roll decreases fruit color, raises the acidity, and delays ripening. The disease is spread by propagation from infected mother vines.

Pierce's Disease (bacterium - rickettsia-like): Early summer symptoms include delayed shoot growth, leaf mottling, and dwarfing of new shoots. Late summer and fall symptoms are burning, scalding, or drying of leaves; wilting or premature coloring of fruit; and uneven cane maturity. Ribier is very susceptible, usually dying within two years. Thompson seedless and most other French (Vinifera) varieties die within two to five years. American (Lambrusca) varieties often live longer than five years. No effective control is known. Pierce's disease is spread by several types of leafhoppers, by the spittlebug, and by grafting. Seventy-three plant species serve as disease reservoirs and hosts for these vectors.

Powdery Mildew (fungus - Uncinula necator): This fungus grows on all above ground parts of the vine. Powdery mildew causes curling and withering of young leaves and dark staining on the surface of mature leaves. It may appear as a gray powdery growth on canes, and when rubbed off, it leaves web-like, dark-brown discolorations. Other symptoms include dropping, discoloration or splitting of berries, browning, and poor maturation of canes. American grape varieties are rarely damaged by this disease. Since the fungus is favored by low humidity and can grow at 90 degrees F., this disease is more common in West Texas. Prevention is the best means to control powdery mildew. It should be controlled by fungicides applied when foliage first develops and repeated at two to three week intervals until berries are full size. If newer fungicides are not available, sulfur dust applied at the rate of 5 to 10 pounds of dusting sulfur per acre will prevent powdery mildew from developing. Apply dust when shoots average 6, 12, and 18 inches in length. Make additional applications every two weeks until fruit matures. Reapply sulfur after rains. If early sprays are omitted, heavy infections can be controlled with wettable sulfur at 1.5 pounds/100 gallons with a wetting agent. Do not spray table grapes with this mixture if berries are more than one-third full size. Sulfur dust or spray can burn leaves, shoots, and fruit when the temperature is over 90 degrees F.; no applications should be made at such times.

DISEASES OF MANGO

Sunburn - (physiological): The surface of a portion of the fruit is dried and slightly sunken, but firm. This is an injury that can result from heating when water or sprayed material on the fruit surface concentrates sunlight.

Tipburn - (physiological): The leaves are necrotic around the edges. A moisture deficiency or salt accumulation can cause the damage. Extreme fluctuations in soil moisture should be avoided. Plants may need a heavy watering to flush salts out of the root zone. This may require an improvement in soil drainage.

DISEASES OF MAYHAW

Brown Rot (fungus - Monilinia sp.): Only the blossom blight phase of brown rot has been observed in Texas and it is of minor importance. Brown rot may become more prevalent as mayhaws are commercialised.

Cedar-Quince Rust (fungus - Gymnosporangium clavipes): Cedar- Quince Rust is the most destructive disease on mayhaw. Infection takes place in spring. Fungal spores for infection on mayhaw come from orange jelly-like growths on cedar limbs - Eastern red cedar primarily. These spores are windblown from up to a mile away, indicating the alternate host for cedar-quince rust does not have to be in close proximity. Mayhaw fruit is more susceptible than young twigs. Leaves are seldom infected. After colonising fruit the fungus erupts through the epidermis forming tubular orange structures full of spores that can only infect cedar or some similar alternate host. Infectious cankers on cedar are perennial. Therefore, yearly re-infection of cedar is not necessary to keep the disease active. Heavily infected mayhaw fruit does not ripen properly and is of no value. Fungicide protection up to four weeks after bloom or until fungal eruptions on cedar cease is essential for control. Fungicides are being developed but are not currently available.

Fire Blight (bacterium - Erwinia amylovora): The bacterium is spread by wind, rain and insects during bloom. Infected blooms die. Invading bacteria progress into tender shoots causing a shoot blight. The bacterium becomes inactive in hot weather. This is a spring and early summer disease that is generally not encountered every year.

DISEASES OF PAPAYA

Anthracnose (fungus - Colletotrichum gloeosporioides): The initial symptoms are watersoaked, sunken spots one-quarter to one inch in diameter on fruit. The centers of these spots later turn black and then pink when the fungus produces spores. The flesh beneath the spots becomes soft and watery, which spreads to the entire fruit. Small, irregular-shaped watersoaked spots on leaves may also be seen. These spots eventually turn brown. In the climate of south Texas, cultural control measures should be sufficient for control of this disease in the home garden. These measures include removing the fruit as soon as it matures, removing all dead leaves and fruit from the vicinity of the plants, and removing infected fruits from the trees. Under conditions of severe disease pressure that would be found in more rainy, humid climates, fungicides may be used.

Black Spot (fungus - Asperisporium caricae): The initial symptoms are water-soaked spots on the upper leaf surface. Later, small, black spots are visible on the underside of the leaf. Black spots may also be found on the fruit. The tissue beneath them becomes corky, but the fruit does not rot. If there is a high incidence of infection of young leaves, a fungicide may be used.

Fire Blight (bacterium - Erwinia amylovora): The bacterium is spread by wind, rain and insects during bloom. Infected blooms die. Invading bacteria progress into tender shoots causing a shoot blight. The bacterium becomes inactive in hot weather. This is a spring and early summer disease that is generally not encountered every year.

Anthracnose (fungus - Colletotrichum gloeosporioides): The initial symptoms are watersoaked, sunken spots one-quarter to one inch in diameter on fruit. The centers of these spots later turn black and then pink when the fungus produces spores. The flesh beneath the spots becomes soft and watery, which spreads to the entire fruit. Small, irregular-shaped watersoaked spots on leaves may also be seen. These spots eventually turn brown. In the climate of south Texas, cultural control measures should be sufficient for control of this disease in the home garden. These measures include removing the fruit as soon as it matures, removing all dead leaves and fruit from the vicinity of the plants, and removing infected fruits from the trees. Under conditions of severe disease pressure that would be found in more rainy, humid climates, fungicides may be used.

Collapse of Mature Plants (unknown cause): This disease has been reported in the Rio Grande Valley. One of the symptoms is a sudden weakening of petioles that causes the leaves to droop close to the stem. The roots or stems of such plants often show a watery brown rot and the plants eventually die. The cause has not been determined, but fungi such as *Rhizoctonia* and *Fusarium* have been isolated from roots. The disease has been associated with waterlogged soil and poor soil drainage.

Damping-Off (fungus - Rhizoctonia solani): This is a disease of young seedlings. Lesions are seen on the stem at or just above soil level. The stem becomes watery and shrinks, followed by death of the plant. Use well-drained soil and avoid overwatering.

Herbicide Injury: Curling of the leaves was seen with the use of a hormonal herbicide (2,4-D) applied to a nearby lawn. Herbicides may drift with the wind from the area of application to nearby susceptible plants. Additionally, ester formulations of 2,4-D may volatilise (i.e. become a vapor after they are applied) and move to non-treated areas. Herbicide injuries of this type are not likely to cause permanent damage to trees.

Stem Rot (fungus - Pythium sp.): This disease was seen on young transplants. The initial symptom is a watersoaking of the lower stem followed by its disintegration. Plants eventually wilt. Often, the white, cottony growth of the fungus is visible. Transplants should be planted in soil with good drainage. Stems should not be banked with soil.

DISEASES OF PEAR

Pyrus communis

Fire Blight (bacterium - Erwinia amylovora): The bacterium causing fire blight overwinters at the margins of the cankers formed on twigs and branches in the previous season. Active bacteria are in the healthy tissue next to the canker rather than in the canker. In the spring the bacteria begin to multiply at the same time growth starts. As the bacterium increases, an ooze is formed at the margin of the canker. Insects are attracted to the ooze and it is carried to the open blossoms. Splashing rain can also spread the bacterium. Once bacteria enter a blossom, the blossoms are blighted within 7-10 days after infection. Bacteria continues to be spread further by insects visiting the blighted blossom and carrying the bacteria to adjoining blossoms. Spread by honey bees is increased during periods of warm temperature, sunshine, and still air. After blossom infection, bacteria spread into the fruit peduncle and finally into the twig. Ooze is continually being produced which can add to secondary infection. During periods of high humidity, the bacteria can enter into young leaves. Fire blight is favored by a mean temperature of 60 degrees or above. The control of fire blight is based on several steps:

1. Plant resistant or tolerant varieties: Kieffer, Orient, Garber, or Douglas.
2. Maintain balanced fertiliser level. Do not use excess levels of nitrogen.
3. Prune during dormant months. Summer pruning may encourage tender succulent growth which is more susceptible to disease.

4. Remove overwintering bacteria cankers by pruning. Make pruning cuts 8 to 12 inches below visible sign of disease.
5. Apply bactericides on 5 day intervals between early bloom and late blooms.
6. Reapply a bactericidal spray to an orchard if it is damaged by hail or receives a heavy rain immediately after an application.

Leaf Blight and Fruit Spot (fungus - Entomosporium maculatum): Leaf spots first appear as small purple spots which enlarge. As the spots get older, they develop purple margins with brown centers. Fruit spots are one-fourth inch in diameter, black, and slightly depressed. They sometimes coalesce to cover a large portion of the fruit surface. Lesions also occur on twigs and are the overwintering inoculum. In the spring, twig lesions produce spores which are washed by rainfall to leaves. After infection there is a period of one week before symptoms are observed. Secondary infection can occur during the spring and summer when the temperature is near 75 degrees F. and surface moisture is on the leaves. Fungicides should be applied at full leaf development and continued at two week intervals for four sprays.

Bitter Rot (fungus - Glomerella cingulata): The fungus attacks apples and pears and enters the fruit through uninjured skin. Infected fruit are characterised by a firm rot which forms a circular light brown spot. With age the spots become almost black and have a saucer-shaped depression. The organism overwinters in decayed fruit and in cracks on the old bark. A broken limb or twig will serve as an overwintering site. Symptoms are first noticeable in mid to late July. A temperature of 85 degrees F and light rain favor development of the rot. Good sanitation will help reduce losses from this disease. Remove all broken limbs and decayed fruit. Once the disease begins to show up, spray at seven day intervals for two or three applications.

Black Rot (fungus - Physalospora obtusa): Black rot is a firm-textured rot. The spot at first is light brown but darkens with age. A circle of raised dark postules are formed in the center of the spot. Infected leaves are covered with many small purple specks. At maturity the spots are purplish cast with brown centers. Twig infections are small, sunken, reddish-brown areas. The organism overwinters in cankers, decayed fruit, and dead wood. In the spring spores are formed in the cankers. Leaves are the first tissue to be infected. A temperature of 80 degrees F. and rainfall encourage disease development. Infection generally takes place at the blossom end of the fruit as it reaches maturity. Sanitation is one of the more important means of control. Remove all dead twigs, limbs, and decayed fruit. Fungicides must be applied when the fruit is beginning to expand in the spring.

Pear Varieties and their Reaction to Fire Blight

Variety Reaction	
Keiffer	R
Orient	R
Garber	MR
Douglas	MR
Bartlett	S

S = Susceptible
MR = Moderately Resistant
R = Resistant

Bot Rot (fungus - Botryosphaeria ribis): The fungus attacks both woody tissues and fruit. On limbs, new infection shows up as small blisters. These lesions serve as a source of inoculum for next spring. Fruit infection results in small, reddish-brown spots which develop rapidly causing a soft rot. A temperature of 75 degrees F or above favors the development of the fungus. Bot rot is most damaging on weak trees. During periods of rapid twig growth, the diseased area will be sloughed off. Prune out any dead or diseased wood. Fungicides will need to be applied from immediately after bloom until near harvest.

Pear Scab (fungus - Venturia pyrina): Symptoms of apple scab and pear scab are similar except pear scab will attack twigs. Twigs are a source of early spring inoculum. The control program for apple scab will also control pear scab.

Stony Pit (virus): This disease is one of the more common viruses found on pears in Texas. The outer flesh of infected fruit is gnarled and deformed. When the fruit is cut, it has brown, hard structures scattered throughout the flesh.

PLUM

Prunus domestica* and *Prunus salicina

Bacterial Canker (bacterium - Pseudomonas syringae): Cankers develop at the base of infected buds on trunk and scaffold limbs. Cankers spread more rapidly above the point of infection than below and only slightly to the sides. This results in a long, narrow canker. Cankers develop during the fall and winter but are not visible until late winter and early spring. Damaged areas are slightly sunken and somewhat darker in color than surrounding bark. As the trees break dormancy in the spring, gum is formed and flows down the outside of the tree. Cankers have a soured smell. Cankers start development in the fall at the base of a bud. The infected areas increase in size during the cool fall months. If infection takes place early in the fall when the trees are still active, the area is walled off by callous and no further-development occurs.

The bacterium is a weak pathogen and causes serious damage only when a tree is in a near dormant condition or weakened due to unfavorable growing conditions. The bacterium is spread by wind, rain and infected budwood which can also be a method of spread. Avoid using high fertiliser rates in the late spring or early summer. This allows a tree to enter dormancy normally rather than encouraging late fall growth which is more easily infected. Prune when trees are fully dormant (January and February). Trees pruned early in the year prior to dormancy can be infected with the bacterium carried on pruning shears. Trees showing signs of bacterial canker should be left and pruned after all other trees have been completed. Early pruning encourages late fall growth of trees which is more susceptible to the bacterium. When pruning diseased trees, dip pruning equipment in a solution of 1 part household bleach and 9 parts water. At the end of the day, rinse and oil lightly to stop corrosion.

Brown Rot (fungus - Monilinia fructicola): This is a fungal disease of plums which attacks fruit in the blossom stage, as the fruit approaches maturity, and during storage. Surface moisture and moderately warm temperatures encourage its development. Fruit damaged by wind, hail, insects, or mechanical means is more susceptible to this organism. Blossom blight is the stage where blossoms are killed by the fungus. Infected blossoms are brown and water-soaked. The fungus grows down the pedicel into the stem which may cause twig dieback or a canker. During high humidity, diseased blossoms and stems may become covered with "tufts" of gray spore masses. The "tufts" are inoculum for further infection as the fruit matures. On mature fruit, the fungus enters through natural openings or wounds and rapidly develops a brown, water-soaked lesion. The organism overwinters in mummies, stem cankers and old fruit peduncles. Control is by repeated fungicide application and sanitation. Picking containers should be cleaned with a captan solution. Packing containers should be new.

Bacterial Spot of Plum (bacterium - Xanthomonas pruni): Symptoms are observed first as small, irregularly shaped lesions. The spots are pale green in contrast to the dark green surrounding tissue. In advanced stages, angular lesions are formed and surrounded by a halo of lighter colored tissue. The inner portion of the lesion turns black and drops out at maturity. This gives the leaf a "ragged" or "shot hole" appearance. Leaves heavily infected with bacterial spot turn yellow and eventually fall. Disease first appears as a small, olive-green, circular spot on the fruit surface. However, as the disease develops, spots become darker and lesions are scattered over the fruit surface. In advanced stages of fruit infection, "pitting" occurs. "Pitting" results when bacteria kill cells in the lesion, and as surrounding healthy tissue grows, a "pit" or split is formed. This "pit" serves as an entry for other disease causing pathogens. During periods of high humidity, gum flows from the lesion which further disfigures the fruit. The area below the "pit" is corky and unpalatable.

Fruit infected with bacterial spot is unsuitable for sale. Twigs are damaged by two distinct types of lesions. "Spring cankers" which develop on young, succulent twigs of the previous summer's growth appear as water-soaked, slightly darkened blisters about the same time as first leaves appear. As the season progresses, the epidermis over the lesion ruptures and releases bacteria. In time the lesion heals and becomes inactive. "Summer cankers" are restricted in size with indefinite margins. Bacteria overwinter in summer canker infected twigs. The summer cankers are the overwintering site for the bacteria.

Bacteria also reproduce in lesions on the leaves and this serves as inoculum for later infection. Repeated infection occurs throughout the growing season as long as the environment is favorable for disease development. In the fall, bacteria are carried to young, succulent stems where infection is through natural openings. If the trees have stopped active growth, bacteria enter the host and undergo limited development. However, if the trees are still actively growing, the host plant responds by walling off bacteria.

This results in the bacteria dying. Chemical control during the season is ineffective. Dormant sprays have been somewhat effective, the spray is accurately timed to protect the stem tissue during the fall infection period.

Rust (fungus - Tranzschelia discolor): Rust occurs on both peach and plum trees. Reddish pustules occur on the lower leaf surface marked by a yellow spot on the upper surface. It causes premature defoliation. After the crop has been harvested, make applications of a protective fungicide on a 2-3 week interval until October 1.

Cherry Leaf Spot (fungus - Coccomyces hiemalis): Infected leaves have small circular purple lesions which turn brown and eventually drop out. This gives the foliage a ragged or shot hole appearance. The fungus overwinters on old leaves. The disease is most active at 60 to 68 degrees F. Rainfall is necessary for disease development. Protective fungicides must be applied throughout the growing season to get effective control.

Black Knot on Plum (fungus - Apiosporina morbosum): Affected limbs and twigs are malformed as a result of black woody growths (galls). The galls are similar to those caused by crown gall on the root system. The disease is more prevalent on the small twigs, but under severe disease conditions it may be found on large limbs. Infection of new wood occurs from bloom until shuck split stage. Primary inoculum comes from one to two year old galls. The fungus can penetrate the uninjured epidermal stem tissue. As the fungus grows, it may develop further down the twig and form a secondary gall. Control of black knot is by sanitation. Prune and destroy the galled tissue. Make cuts two or three inches below the gall to insure complete removal of the fungus. Fungicides applied in the early bloom periods will help reduce the occurrence of this disease.

Plum Pockets (fungus - Taphrina pruni): Fruit infected with the plum pockets fungus are puffy and distorted and they fall soon after infection. The first symptom of infection is presence of small whitish spots on the plum. This happens two to three weeks after shuck split. Once infected, the flesh becomes spongy and the embroyo is killed. The skin of the fruit is first a reddish color and then a velvety gray. For this disease to develop, cool, wet conditions are necessary during early fruit development. It is similar to peach leaf curl in its dependence on weather for infection. The fungicide program used to control peach leaf curl will also control plum pockets.

STRAWBERRY

Fragaria chiloensis* and *Fragaria ananassa

Leaf Spot (fungus - Cercospora fragariae): This is one of the more common leaf spots of strawberry. It occurs in all areas of Texas. At first, small round purple spots are seen on the upper side of the leaves. As the leaf matures, the center of the spot becomes tan or gray then almost white while the edges of the spot remain purple. On the underside of the leaf, the spots are bluish or tan in color. The fungus causing leaf spot overwinters easily in Texas since strawberries are a perennial plant in Texas, and there are always present enough infected leaves to provide inoculum. Varieties vary in their resistance to leaf spots. Chemical control can be obtained using a well planned spray program with repeated application. Sprays must be applied with enough pressure to obtain thorough coverage on the upper and lower side of the leaves.

Leaf Scorch (fungus - Marssonina fragariae): The disease is similar to leaf spot. At first, small dark purple spots up to one-fourth of an inch in diameter appear on the upper side of the leaves. These spots never have light centers as do those of leaf spot and they have a more irregular outline. At maturity, the leaf spots are large and irregular in outline and never show the white center described for leaf spot. The disease can also attack other parts of the plant, and the spots may appear on other affected parts. Sanitary measures such as raking and burning of old leaves, clean cultivation, and frequent renewal of the plants may help reduce disease losses. Some varieties have resistance to leaf scorch. Fungicide applications at repeated intervals should give adequate control. Follow recommendations as for leaf spot control.

Red Stele (fungus - Phytophthora fragariae): This is the most serious root disease of strawberries. It occurs during the late winter and spring. This fungus attacks only strawberries. The fungus is spread from one area to another primarily through diseased plants. In a healthy plant, the roots are a yellowish-white and in a diseased plant, the center is a distinctive brownish-red. Infected plants are stunted and wilt in dry weather. The root systems have a rattail appearance with few lateral roots produced. The disease is more disastrous in low, poorly drained fields during cool weather. There is no chemical or cultural methods for reducing disease losses, except to replant resistant varieties. The fungus does not persist in well drained soils. To prevent the possible build-up of a more virulent strain of the fungus, crop rotation should be practiced. Transplants should be thoroughly examined for the rattail appearance to prevent bringing in diseased plants. Avoid planting any but resistant varieties which are adapted to your area of Texas if the disease is known to occur in your garden or field.

Verticillium Wilt (fungus - Verticillium albo-atrum): This is a wide-spread soil-borne disease of strawberries occurring in cool, poorly drained soils. Symptoms show up as a burning of the outer margins of the leaves. Young roots turn black at the tips and die back. If Verticillium has been a problem in your area, consider a variety which has resistance to the wilt organism. Soil fumigation with a chloropicrin-containing fumigant is recommended when resistant varieties are not available. When using soil fumigants, follow manufacturer's directions and use caution in handling the material.

Black Root Rot (various fungi): The exact cause of black root rot is unknown. It has been associated with several different organisms or conditions. The symptoms of black root rot are a black and woody root system. On healthy plants, the roots will appear white and pliable; however, diseased plants will have a stiff root system. The feeder roots are partially or completely destroyed. The tip of the main roots will be destroyed. Fumigation has helped with this problem.

Anthracnose (fungus - Colletotrichum fragariae): The fungus attacks leaves, stolons, petioles, fruit and plant crown. Leaf infection is known as black leaf spot. Elongated brown to black lesions appear on stolons and petioles when infected. The most serious phase is crown infection which results in poor runner production and plant death. A longitudinal cut through the crown reveals a firm, reddish brown rot. Avoid excess fertility which favors disease development. Anthracnose is difficult to control with fungicides. When observed use fungicides as often as label allows.

Fruit Rots (fungi - Botrytis cinerea, Rhizoctonia spp.): fruit rots destroy a large percentage of the fruit each year in Texas. There are several organisms involved in this complex. Symptoms will be somewhat different for each disease. Gray mold rot Botrytis) is the most damaging. It attacks the green fruit as well as the ripe fruit. The first sign of infection is where the fruit has touched the ground or another diseased strawberry. It begins as a small, brown, soft spot. the spot spreads rapidly, often giving the berry a whiskered effect. To control this disease, the plants must be spaced so that the foliage or fruit does not stay wet for extended periods of time. The hard rot, caused by Rhizoctonia, attacks only ripe fruit that are in contact with the soil while growing. Affected berries are usually one sided and show a hard brown area on the side that touched the ground. Other control measures that will help reduce diseases should be used. Avoid applying heavy applications of fertiliser in the spring. Spring fertilisation will cause the foliage to be dense and thus reduce the rate at which the soil dries. Spraying or dusting with fungicides will prevent harvest rots if properly applied. Spray with a fungicide as soon as the young fruit bud becomes visible and continue spraying every two weeks until the fruit is one-third grown. Additional sprays may be required if the weather is cool and wet.

REFERENCES

Fortune M and Dilbar A., "New diseases of mango (Mangifera indica), passion fruit (Passiflora edulis var flavicarpa) and pommecythere (Spondias cytherea) and their threat to sustainable agriculture", *Agriculture in a changing economic environment: Proceedings of the Seventh Annual Seminar on Agricultural Research*, San Fernando, Trinidad and Tobago, 9-10 November 1993.

Jeffries P and Jeger M J., "The biological control of postharvest diseases of fruit", *Biocontrol News and Information*, 1990.

Whitwell A., "Dominica orchard crop management and research project: pest and disease management", *Final report: minor Tropical fruit crops*, Kent, United Kingdom: Natural Resources Institute, 1991.

6

Diseases of Nut Crops

Like the small fruits, nut crops are a group of botanically unrelated crops. They are grouped together because the fruit type is a nut (generally), and harvesting, post-harvest processing, nutritional value and marketing characteristics are relatively similar. There are four major temperate-zone nuts, pecan, walnut, almond, and pistachio, and ended up with two tropicals - cashew and macadamia.

Nuts are often wind pollinated, unlike many of the species we have considered thus far. The trees generally produce a lot of pollen, often in specialised male inflorescences, to ensure that the female flowers will be fertilised. While possible, we would not want parthenocarpic fruit set in nut crops, since the seeds are the edible products.

The terms monoecious and dioecious are useful in this section, as pecans and walnuts are all monoecious, and the pistachio is dioecious. Take a look at these definitions in the glossary. Also look at "protandrous" and "protogynous" under "heterodichogamy", and think about the implications for orchard design and polliniser placement.

Walnut and pecan are closely related, both members of the Juglandaceae. While they are not cross- or graft-compatible, there are several aspects of their botany that are similar. One of the main differences, responsible for much higher yields in walnuts than pecans, is the propensity to bear nuts from lateral buds on 1-yr-old wood in walnut, not just terminal buds. This one feature makes a big difference in the yields per acre and orchard design between pecan and walnut.

Nut crops often undergo a process known as alternate bearing, or irregular bearing. In other words, yield is high one year then low the next. Reasons for this revolve around two main theories: a hormonal theory and a carbohydrate depletion theory. The hormonal theory states that hormones in the developing nuts diffuse back into the stems and inhibit flower bud formation for next year's crop. The carbohydrate theory states that carbohydrate reserves are exhausted in high crop years, such that none are left to allow flower bud formation for the next year. In "off" years, either the lack of hormone inhibitors or the surplus of carbohydrates available allow for great flower bud set and a big crop the following year. In pistachio, it works a bit different, where flower buds are set for next year's crop in "on" years, but fall off during the maturation of the current season's crop. Regardless of the mechanism, this creates a hardship on the grower, since yields, and therefore prices, fluctuate drastically from year to year. Some progress has been made with "thinning" nutlets in an "on" year in pecan, which has the effect of smoothing out the variation in yield over time.

The pecan is the most important native orchard species in North America. Other than some chestnuts, none of the other nut crops are native. Notice the lower yields per acre of nut crops compared to fresh fruits, due to the fact that nuts are dried to about 5-10% water content before sale, while fresh fruits are sold at about 90% water content. For example, 1-2 tons per acre would be a good yield for nuts, but yields of 10-20 tons/acre are typical in tree fruits. Despite lower yields, profits in nuts can be quite high since the price per lb is often over $1, while the price per lb of fresh fruits is generally 10-30 cents/lb.

Nuts are generally harvested by shake-and-catch methods, where the trunk is gripped by a mechanical arm, and shaken for a few seconds to dislodge the nuts. Nuts are then swept from the ground or caught in "catch frames" and funneled into storage bins. Few fresh fruits could be harvested this way since they would be bruised or cut as they fell, rendering them unmarketable. Harvesting costs are substantially lower (per lb) in nuts, despite a high initial investment in a mechanical harvester.

Pecans and walnuts are the largest orchard species grown, and it is common to find planting densities of only 10-40 trees/acre, as opposed to 100 to several hundred plants/acre in tree and small fruits.

In pistachio, almond, and macadamia, the fruit starts out as a drupe, but is marketed as a nut; review the definitions and structures of drupes and nuts and try to figure out what makes this possible.

DISEASES OF BLACK WALNUT

Chestnut Blight

The symptoms on Chinese chestnuts are cankered areas on the bark of a branch or the trunk. New cankers show sunken or swollen areas of bark; later, the bark may split and the foliage may wilt.

Good cultural practices and resistant varieties are the best means of control. Keep the trees growing with normal vigor and prune out cankered branches.

Pecan Scab

Scab is a major disease on several southern pecan varieties, but northern pecan varieties, hickories, and hicans are usually not severely attacked.

Symptoms of scab are round or irregular olive-brown to black spots on leaves and young twigs and small, dark, circular spots on the husks of the plants.

For home plantings, control with good sanitation. Rake up and burn or haul away all hickory, pecan, and hican leaves, shucks, and dead twigs. Where scab, leaf blotch, leaf scorch, spot anthracnose, anthracnose, or other fungus leaf spots are serious, apply 4 to 6 sprays 10 to 14 days apart. Start when the buds begin to open. Suggested fungicides include benomyl, maneb, mancozeb, and dodine. Follow label directions.

Walnut Anthracnose

Anthracnose is the most serious disease of black walnuts, although Persian walnuts are resistant to it. Anthracnose attacks leaves, nuts, and new shoots. Wet weather during late spring and early summer increases the severity of the infection. Severe infection causes leaves to drop prematurely, sometimes partially defoliating the trees by midsummer.

Starting in May or June, small, dark spots appear on the leaves. These spots enlarge and may merge to form dead areas. Tiny, sunken, dark spots develop on the husks of the nuts. Husk infection early in the summer may cause the nuts to drop prematurely or to be improperly filled. Defoliation of the trees may also result in improperly filled nuts with dark kernels.

Sanitation and using the more resistant varieties are suggested for control. Each spring, rake up and burn or haul away all walnut leaves. The varieties in the suggested list show resistance to anthracnose.

Where anthracnose, yellow leaf blotch, and other fungus leaf spots and blights are serious, start spraying when the leaves begin to unfold and continue at 2-week intervals 3 or 4 times. Suggested fungicides include benomyl, dodine, maneb, and mancozeb.

Walnut Blight

Persian walnuts are more susceptible than black walnuts to

this bacterial disease. Blight attacks leaves, bark, shoots, and nuts. Infections on new shoots do not grow into older wood, so trees are not killed, but the nuts can be severely damaged and fail to fill properly. Nuts may be infected anytime during the growing season. First symptoms are small, water soaked spots on the nuts, leaves, or shoots. These spots enlarge and become dark and sunken.

Use fixed copper (50 to 55 percent copper) at the rate of 4 pounds per 100 gallons of water (3 level tablespoons per gallon). Spray 3 times: at the beginning and completion of flowering and at nut set.

Walnut Bunch Disease

Bunch disease affects black walnuts, pecans, and hickories and is especially serious on butternuts, Japanese heartnuts, and hybrids of butternuts and Japanese heartnuts. The causative organism and the method of transmission are not known, but some scientists suspect that a virus or a mycoplasma is involved.

Lateral (side) buds have a tendency to grow rather than remain dormant. This produces a "witch's broom" type of growth on the infected branches, characterised by bushy, closely spaced lateral shoots with undersized leaves. Upright, suckerlike shoots form on the trunks and main branches.

Infected branches frequently start growth earlier than normal in the spring and grow longer into the fall. This late fall growth retards normal cold-hardiness development, and the tips of infected branches are winterkilled. Branches infected with bunch disease do not produce normal crops of nuts.

For control, cut out the infected branches, making the cut well below the infected area. If the disease continues to spread into other branches, remove the entire tree.

PECAN

Carya illinoensis

Scab (fungus - Cladosporium effusum): The fungus invades growing shoots, leaves and later developing nuts. Severely infected nuts fall or fail to fully develop. Infected foliage prematurely drops during periods of frequent rainfall and mild temperatures. These conditions are optimum for rapid development and spread of the scab fungus. Infection takes place when leaves or nuts are covered with free moisture for at least 2 hours. The scab fungus overwinters in infected shoots, old shucks, and leaf petioles. In the spring when temperature and moisture conditions become favorable, the fungus becomes active and within a few days spores are formed at the surface. They are spread by wind and rain to developing leaves and nutlets. If weather conditions are favorable, the spores germinate and enter the host tissue. Within 7-10 days new lesions are formed on the host.

Lesions formed on the lower leaf surface are characteristically olive brown in color, somewhat elongated in shape. They range in size from a barely discernible dot to lesions one-fourth inch or more in diameter. Frequently, adjacent lesions coalesce, forming large, chocolate brown irregularly shaped spots. Scab lesions occur on or along the leaflet veins, but can be found between the veins. On nuts, the scab lesions are small black dots which become sunken as they mature. Adjacent lesions on the nuts may coalesce, forming large sunken black lesions. When infection is severe, the entire nut surface is black, development is arrested and the nut drops prematurely or fails to grow in the area of infection.

Pecan varieties vary in their susceptibility to the scab fungus. Strains of the pecan scab fungus can develop that will infect a resistant variety. As the inoculum from these strains increase, a once resistant variety is increasingly infected by the fungus. Even though a variety possibly will not be resistant throughout its entire life, it is best to plant varieties that are productive and resistant to the scab fungus.

Trees should be spaced to allow for air circulation between the trees. This encourages rapid drying of the foliage. Properly spaced trees will have a higher percentage of their leaves exposed to sunlight. Removal of the lower limbs of a tree will aid in air circulation and reduce the time leaves remain wet following rain or heavy dews.

Trash trees surrounding the field should be removed. Pecan trees that can not be adequately sprayed should be removed if they are susceptible to the scab fungus. Spores of the pecan scab fungus produced on the unsprayed trees are spread by air currents to surrounding trees. Sanitation measures such as removal of old shucks and leaf petioles from around trees wiil help to reduce disease inoculum in the orchard. If left in place, spores produced from old lesions are carried by air currents up to the tree.

Fungicides are an important part of the pecan disease management program. Although some of the newer fungicides have post infection activity, they are most effective when applied as a preventive application. Thorough coverage of leaf, nut and shoot surfaces with a fungicide must be maintained to prevent infections.

Brown Leaf Spot (fungus - Cercospora fusca): Brown leaf spot affects only mature leaves and does not appear until the latter part of May or mid-June. Primary lesions develop on the lower leaf surfaces as small dots which gradually enlarge and become reddish-brown with a gray cast. The shape of the lesions may be circular or irregular, especially where two or more lesions develop adjacent to each another. In seasons favorable for brown leaf spot development, pecan trees may be completely defoliated if the fungus is not controlled. Only the foliage is infected by this fungus. Pecan trees that are making active growth are less likely to have brown spot. Leaves on trees that are fertilised properly sedlom are infected with the brown leaf spot fungus. The problem is most often observed on pecans growing in sandy soil. Heavy rains during the spring will leach nitrogen below the root zone of the trees. Fungicides applied as preventative sprays will protect the foliage.

Vein Spot (fungus -Gnomonia nerviseda): Foliage symptoms of this disease are similar to the lesions formed by the scab fungus. Unlike scab, vein spot only affects the leaves. Lesions of vein spot disease develop on veins of leaflets and are characteristically dark brown to black. Severely infected leaflets shed soon after symptoms are observed. The fungus overwinters in fallen leaves beneath the tree. Fungicides and sanitation procedures will control this fungus. Fungicides applied for scab control are also effective against the vein spot fungus.

Leaf Blotch (fungus - Mycosphaerella dendroides): This disease occurs mainly on low vigor trees. It overwinters in infected leaves on the orchard floor. Symptoms first appear on the under surface of mature leaves in early summer as small olive-green velvety spots. By midsummer, black pimple-like dots are visible in the spots. When severe, infected leaflets are killed, causing leaf shed in late summer or early fall. Premature shedding adds to the reduced plant vigor. Fungicides applied following the pecan scab spray program will also control this fungus.

Downy Spot (fungus - Mycosphaerella caryigena): Foliage on trees infected with this fungus shed soon after symptoms are observed. Primary infection occurs on developing leaves in the spring from spores produced on last years foliage. Symptoms usually appear during the summer months on the lower surfaces of leaflets. Lesions are one-eighth to one-fourth inch in diameter and greenish-yellow. Later in the season, the lesions turn brown due to the death of the leaf cells in the disease area. Spores are spread by wind and rain to nearby leaves and neighboring trees. Although all pecan varieties are moderately to slightly susceptible, Pawnee, Moneymaker and Stuart varieties are the most susceptible to the downy spot fungus. Disk or rake under fallen leaves in the fall or early spring before the leafbuds begin to swell. By removing the leaves from the soil surface, spore discharge is stopped. Fungicide applications between bud break and casebearer are the most critical in the control of downy spot.

Powdery Mildew (fungus - Microsphaera alni): Infected nuts are covered with a white powdery substance. The fungus most often develops between mid-July and mid-August. Due to rainfall washing the spores from the nuts or leaves, the fungus seldom is a problem during months of frequent rain. Pecans which were previously infected by the powdery mildew fungus are brown in color when the fungus is washed by rainfall. The fungus is sometimes associated with nursery trees or sucker growth around the base of mature trees. Infected leaves are somewhat distorted in shape and are covered with a faint gray colored dusty material.

Powdery mildew infected nuts or leaves are not significantly damaged by the fungus. Its feeding is limited to the outer cells of the shuck. This does not restrict the movement of nutrients into the developing kernel.

Fungicides applied in the course of the regular scab spray program will control powdery mildew.

Fungal Leaf Scorch (fungus - several): Infected leaves turn brown along the leaf margin or tip. The necrotic areas may extend from only a small area at the leaf margin to completely encircle the margin of the leaf. As the lesion matures it becomes gray. Infection can result in leaf shedding from mid July to mid August. Symptoms develop on a few leaves and then spreads to nearby leaves. Seldom is the entire tree affected. Because the infected leaves shed, orchard insepections made late in the season are not effective in diagnosing the disease. Several fungi are reported to be the cause of this disease. All of the fungi are controlled with approved fungicides. Fungicide applications applied for scab are effective when applied in early to mid summer. The varieties Western, Cheyenne, Chickasaw, Shoshoni, Stuart, and Cape Fear are some of the more susceptible varieties to this group of fungi.

Bunch Disease (mycoplasma): Although the cause of bunch disease is not completely understood, evidence indicates it is a mycoplasma (MLO). Trees affected with bunch disease show a typical bunching symptom, caused by excessive growth of lateral buds. Normally only the most distant bud will break dormancy and develop into a twig and finally a limb. A large percentage of the buds on bunch infected trees will break dormancy at one time. This results in a dense growth of small twigs. In some cases only one or two limbs will be infected by the organism. Nut production is very limited on bunch infected trees. Nuts from infected trees are restricted in size. There is no effective control for bunch disease, except tree removal.

Articularia Leaf Mold (fungus - Articularia quercina): The disease occurs most commonly following rainy periods, in areas of high relative humidity and on leaves of poor vigor trees. The fungus grows on the lower surface of the leaves, a growth of white tufts that contain spore masses. Articularia leaf mold does not occur on trees in orchards that have been sprayed regularly for disease control. A single fungicide application at the first sign of disease is usually sufficient to control articularia leaf mold disease. The materials applied for scab control will also provide some protection against this fungus.

Pink Mold (fungus - Trichothecium roseum): Pink mold infection occurs on scab infected nuts. Infected nuts develop a white mold growth in old scab lesions. In areas not exposed to air and sunlight, the white area becomes pink in color. The fungus may invade the kernel. The kernel becomes oily in appearance and 'leaks' onto nearby nuts.

The kernel is rancid and not eatable. In areas where scab disease control is regularly practiced, pink mold is not a problem. If pink mold is a severe problem, it maybe necessary to remove the oily nuts or to segarate the problem load from disease free nuts.

Cotton Root Rot (fungus - Phymatotrichum omnivorum): Cotton root rot is caused by a soil-inhabiting fungus pathogen that attacks a wide range of host plants including pecan. Roots of pecan tree can be invaded at any time except during the coldest months of the year. The fungus is most active during the mid to late summer months. Infected roots are killed, disrupting the transportation of water to the leaves. Diseased trees die soon after becoming infected. Losses continue several years after planting. The losses increase in numbers roughly in porportion to the increasing size of the disease center. The fungus is present in many alkaline soils in central and south Texas. An effective control for cotton root rot has not been developed. Do not plant in soils that have had a history of cotton root rot.

Shuck Die Back (physiological - Possible hormone imbalance): This condition is commonly associated with Success and Success hybrids. Nuts affected by this disorder drop from one to two weeks early. They do not fill properly due to the peduncle being girdled,and restricts movement of nutrients into the pecan. This results in what is known as "pops." The shuck turns black and opens at the tip in a normal manner. No control is recommended.

Kernel Rot (fungus- Phytophthora sp.): This has been described from the Southeast but has not been confirmed in Texas. It is most often associated with very wet conditions in the later stages of nut development. The shucks turn black. The kernel becomes discolored,and is not eatable. In the southeast copper based fungicides are applied from the water stage to shuck split on a 10 day schedule.

Stem End Blight (fungus - Botryosphaeria ribis): This disorder is associated with a fungus which attacks the nuts in the latter part of July and August. Shucks turn black rapidly and drop soon after. Lesions are black, sunken, and shiny. When infected nuts are cut open, the liquid in the kernel has turned brown. This can be controlled with foliar sprays of a fungicide applied when the pecans begin to fill.

Root Knot on Pecans (nematodes - Meloidogyne sp.) : Small swellings are found on the rootlets. Infected trees develop severe zinc deficiency symptoms. Growers should examine all nursery trees before planting. Chemical control is not recommended.

Kernel Discoloration (fungus - Nematospora spp.): Most kernel discoloration is characterized by distinct dark spot on the kernel. This condition most frequently develops after the kernel is damaged by stink bug attack which is followed by fungal invasion. Improper drying of the pecan can also result in kernel decay.

Fungal Twig Die Back (fungus - Botryodiplodia): Infected twigs are covered with small raised pustules with black centers. This results in twig dieback. Maintain trees in a healthy condition. No control suggested at this time

MAJOR INSECT PESTS

Pecan weevil

Adult weevils feed on young pecan nuts. This feeding injury may cause the nut to drop. Later in the season, eggs are laid in the husks of both pecan and hickory nuts. The eggs hatch into legless grubs, which bore into the nut and eat the kernel. Because the pecan weevil overwinters in the ground, sanitation is not an effective control against this pest, sprays must be applied every 2 to 3 weeks throughout the summer.

To be effective against this pest, sprays must be applied every 2 to 3 weeks throughout the summer.

Hickory Shuckworm

The adult stage of this insect is a moth that lays eggs on or near pecan and hickory nuts. The eggs hatch into caterpillars that feed on the nut or husks. During the growing season more than 1 generation may develop. Before the shell hardens, the caterpillars enter the nut and feed on the young kernel, thus causing the nut to drop. After the shell hardens, the caterpillars bore into the husks. This feeding interferes with normal nut filling and causes black stains on the shell.

The hickory shuckworm overwinters in the husks on the ground. Sanitation will help control this pest in small, isolated plantings. Each fall or early spring, rake up and burn or haul away all husks. During the growing season, burn or haul away any immature nuts that fall.

Walnut huskfly maggot

The maggot overwinters in the ground and emerges in mid- to late summer as a fly. The adult fly lays eggs in the husks of black and Persian walnuts.

The eggs hatch into small legless maggots that feed on the husks. Feeding damage stains the shell and may cause discolored and off-flavored kernels.

Damage usually is more severe on Persian walnuts than on black walnuts. Sprays of carbaryl (Sevin), malathion, or carbaryl and malathion should be applied every 2 to 3 weeks during July and August for control.

Pecan nut casebearer

These insects overwinter as caterpillars enclosed in cocoons attached to the twigs of pecan, hickory, black walnut, and Persian walnut. At bud break in spring, the caterpillars emerge and feed on the buds, then bore into the new shoots. After feeding for a short time, they pupate and change into moths. The moths start egglaying about the time the new nuts start to form. The eggs hatch into caterpillars that feed on the new nuts, destroying them. In Illinois there are usually 1 or 2 succeeding generations that feed on the leaves and shucks.

Injury is more severe on pecans. Control with a spray of carbaryl (Sevin) when the tips of the young nuts turn brown.

Foliage-eating Caterpillars and Worms

Fall webworms, walnut Datana, eastern tent caterpillars, hickory tussock moths, walnut caterpillars, and other leaf-eating caterpillars and worms frequently attack nut trees. When infestations occur, control with carbaryl (Sevin) or Bacillus thuringensis (many).

Plant Bugs, Stinkbugs, and Spittlebugs

These juice-sucking insects damage the nuts. Feeding on young nuts may cause the nuts to drop or may cause part or all of the kernel to turn brown and shrivel.

These pests are general feeders and attack a wide variety of plants, moving about rapidly. They are more prevalent in areas where tall grass and weeds are growing. Keep the nut grove mowed.

CONTROLLING INSECTS AND DISEASES

Many insect and disease pests attacking nut trees can be controlled by spray programs, but the selection of pesticides and timing of sprays must be tailored for the type of tree and the specific pests. Detailed pest control information is beyond the scope of this circular.

Powerful equipment is needed to spray large trees adequately. For the noncommercial grower with a limited number of trees, spraying probably is not practical, except for young trees.

Occasionally, foliage diseases or heavy infestations of insects threaten to defoliate the trees. If the trees are young, spraying is suggested. For aphids, malathion at the rate recommended on the label. For worms and caterpillars, use carbaryl (Sevin) or *Bacillus thuringensis* at the rate recommended on the label.

For fungus diseases of the foliage, such as pecan scab and walnut anthracnose, use benomyl, maneb, or mancozeb at the rates recommended on the label. Maneb, mancozeb, benomyl, carbaryl, and malathion are compatible and can be mixed together in any combination.

Good cultural and sanitation practices will help reduce losses from some insects and diseases. Keep trees growing with moderate vigor. In the fall or early spring, rake up and burn or haul away old leaves, hulls, unharvested nuts, and dead twigs. During the growing season, pick up and burn or haul away any nuts that fall prematurely—they usually have worms in them.

Controlling Nut Predators

In much of Illinois, squirrels and birds are a greater deterrent to successful nut production than plant diseases and insects. In some locations, tree squirrels are numerous enough to take most of the filbert, pecan, hickory, and Persian walnuts before the grower can harvest them at the right stage of maturity. Birds, particularly jays and crows, will harvest filberts and Persian walnuts. Blackbirds and starlings sometimes destroy green nuts of Persian walnuts. Ground squirrels will climb the bushes to harvest filberts and can take pecans and other thin-shelled nuts from the ground. Chestnuts, at least, are protected by their burs until the nuts fall out.

It is well to look at the local squirrel and bird populations and size up their possible effects before putting much time and effort into planting susceptible nut tree species. Chestnuts, hickories, and black walnuts will usually be safer than other kinds.

Losses to tree squirrels can be reduced by using a smooth metal shield placed around the trunk to prevent squirrels from climbing the trunk. This will work if the nut tree is isolated so that squirrels cannot jump from other trees, buildings, or wires to the nut tree, and if the lowest branches are too high for the squirrels to jump to them from the ground. The shield should be 24 inches wide and placed on the trunk at breast height.

Hastening the falling of mature nuts by shaking limbs and frequently picking up fallen mature nuts will reduce losses.

REFERENCES

George Agrios, *Plant Pathology*, New York: Academic Press, 1969.

Jerral D. Johnson and David N. Appel, *Major Oak Diseases and Their Control*, College Station: Texas Agricultural Extension Service, 1984.

Wendell Horne C. et al., *Texas Plant Disease Handbook*, College Station: Texas Agricultural Extension Service, 1983.

7

Diseases Affecting Flowering Plants

AFRICAN VIOLET (SAINTPAULIA)

Saintpaulia ionantha

Crown Rot (fungi - Pythium spp., Phytophthora spp., Fusarium spp.): Crown rot is probably the most serious disease of African violets and may cause loss of entire groups of plants. Older leaves droop and younger leaves showing stunting. Roots are killed rapidly and appear brown. Unless treatment is administered before massive root death, the plant will have to grow an entirely new root system before recovery occurs. Two alternatives are available for infected plants: One is to discard all affected and exposed plants and the other is to use fungicide drenches. Drenches should be administered when the first evidence of disease occurs. Caution should be used in getting the right dosage levels since excessive levels of some chemicals may damage plants. Preventive measures include using sterilised soil and avoiding plant introductions that may harbor crown rot organisms.

Botrytis Blight (fungus - Botrytis cinerea): Leaves, flowers and petioles develop small water-soaked spots that enlarge rapidly. A grayish fungal growth may be seen upon close examination of diseased tissue. The disease is more severe when atmospheric conditions are cool and damp, poor air circulation exists and when light intensity is low. All disease tissue should be removed as this serves as a source for new spores. Surface sterilise the area surrounding the plants with a household cleaner or bleach.

Powdery Mildew (fungus - Oidium spp.): A white powdery type substance may be observed on leaves, petioles, petals or flower stems. It is more easily seen on dark blossoms than white ones even though the white varieties may be slightly more susceptible. Spores of the fungus are air-borne from one plant to another. Control with fungicides is very effective. If the systemic fungicide is used, one may have to spray blossoms since the fungicide is not translocated to floral parts when the pot drench method is employed.

Petiole Rot (Physiogenic): An orange to brown, rust colored spot appears where the petiole touches the pot. The petiole and leaf may collapse. This damage results from salt accumulation on the rim of the pot and the soil surface. Use rain water or another salt-free source of water and avoid over fertilisation. Construct a collar from aluminum foil to be fitted around the rim of the pot.

Ring Spot (Physiogenic): Light brown rings form on leaves with some running together to form irregularly shaped spots. This condition is caused by cold water coming in contact with the leaves. In some cases, damage may occur even if warm water is used. Such a possibility exists when a breeze blows across the wet area and produces an evaporative effect. Water should be kept off leaves.

Root Knot (nematode - Meloidogyne spp.): Galls form on roots of plants causing the root system to be inefficient in absorbing and translocating water and nutrients. This condition can be prevented by using nematode free propagating stock and a sterilised potting medium. The nematode may also be transmitted by petiole cuttings from infected plants. Do not use infected plants for propagation.

Crown Gall (bacterium - Agrobacterium tumefaciens): Fleshy galls form around the base of the plant with a profusion of leaves being produced at that point. Natural infection seems to be slight but infected tissue is perpetuated by some who sell this unusual looking plant under the name of "Witchcraft". Infected plants should be discarded.

Viruses (virus - several): Infected plants have distorted leaves that are mottled in color. Infected plants are not killed but should be discarded because they may serve as a possible source of infection for other plants.

AMARYLLIS

Amaryllis belladonna

Leaf Scorch or Red Blotch (fungus - Stagonospora curtisii): Flower stalks are affected resulting in their being distorted and stunted. Affected stalks have bright red lesions that are one-fourth to one-half inches wide and several inches long. Affected flower stalks grow at an angle instead of the normal erect manner. In the later stages of infection, a white to brownish-gray mycelium develops in the center of cankers. The fungus is carried over in bulbs and discolored bulbs should be discarded. Foliar fungicides may be used to protect above-ground parts.

Bulb Rots (fungi - Botrytis spp., Rhizopus spp., Sclerotinia spp., Pythium spp. and others): Bulb rots caused by fungi are quite common. The fungi may enter the outer scales and progress inward until all scales are infected. Bulbs showing signs of infection should be discarded to prevent contamination of the beds. Soil sterilisation with chemicals or with steam may also be helpful.

Mosaic (virus): Infected plants are stunted and have yellow mottled leaves. The mottle becomes more severe with stunting of all plant parts as time progresses. There is no chemical control and once a plant becomes infected it will remain that way throughout its life. Infected plants should be destroyed.

Southern Blight (fungus - Sclerotium rolfsii): A white fungal growth appears at the ground line and infects the plant at that point. Damage results in death of the above-ground portions. Small bead-like structures (sclerotia) appear among the white fungal strands and serve as overwintering bodies for the fungus. A drench with a soil fungicide is usually sufficient to correct the problem. Avoid the accumulation of plant residue around the plant base

ASTER

***Aster* spp.**

Leaf Spots (fungus - Aschochyta spp., Septoria spp:, Cercospora spp., Alternaria spp., Phyllosticta spp.): Leaf spots may form in any stage of growth. Flower parts and stems may also be affected. Older leaves are generally affected first. Foliage may gradually wither and die starting with the lower leaves. Overhead sprinkling or watering will increase severity of disease. Spores of the causal organisms may be seed borne and use of seed treatment fungicides will reduce disease incidence. Planting in wet, poorly drained areas should be avoided. Foliar fungicides should be applied at the first sign of disease and continued at 7-10 day intervals.

Foot Rot (fungus - Phytophthora cryptogea): A rapid, permanent wilting of the leaves and a brownish-black discoloration appears on the lower portion of the stem. The stem may shrivel and collapse at or near the ground line. Roots may be affected by a soft, water-soaked decay. Adventitious roots may form close to diseased portions of the roots. Diseased plants are pulled easily from the soil, but outer bark of the roots sloughs off and remains in the soil. Plant only in well drained soils. This disease is soilborne and once the soil is infested, asters should not be replanted.

Root and Stem Rot (fungus- Rhizoctonia solani): The fungus may attack roots, stems and lower leaves. Symptoms on lower leaves first appear on those leaves in contact with the soil. Affected leaves are characterised by a well defined water-soaked area that progresses rapidly during cool, moist conditions. At first, the leaf is a dark green; it then wilts and turns dark brown. The stem and root may also be attacked directly. The infected area is first water-soaked in appearance; it later becomes sunken and turns a dark reddish-brown. Control is best obtained by planting only in sterilised beds. Beds should be well drained. Soil fungicides applied as a drench around infected plants may be effective.

Wilt (fungus - Fusarium oxysporum f.spp. callistephi): Both young and old plants may be affected. Young plants wilt and die rapidly. Older plants may exhibit two types of symptoms. One side of a plant may be stunted with a yellowing of the lower leaves on that side. When the stem is cut, a brown discoloration of the vascular tissue is found on the infected side of the plant. In the second type of symptom, lower leaves are yellow and wilt. This is followed by wilting and death of the entire plant. Generally this occurs when the plant is in full bloom. The lower portion of the stem may be dark brown when the stem is cut. Most effective means of control is with the use of wilt resistant varieties such as Gem mixed, or Dwarf Queen.

Powdery Mildew (fungus - Erysiphe cichoracearum): This disease is characterised by a white, powdery growth on the upper leaf surface. Older leaves may be affected with little distortion but young leaves become twisted and misshapen. Young growing portions of the plant may be completely infected resulting in dwarfing and curling of the leaves, stems and buds. Often infected buds fail to open. Effective control requires spraying plants with a recommended fungicide at the first sign of disease.

BEGONIA

***Begonia* spp.**

Stem and Root Rot (fungi - Pythium spp., Rhizoctonia spp. and others): Affected plants wilt and break over at the soil line due to decayed spots being formed on stems. This problem is often severe when cuttings are placed in beds for rooting. Fungi responsible for this condition may be introduced on cuttings or be present in soil. Use sterilised soil. Fungicides may be used as a drench. Use cuttings from healthy plants only.

Botrytis Blight and Stem Rot (fungus - Botrytis cinerea): This fungus is most severe when temperatures are cool and moisture levels are high. Affected plants decline rapidly with stems and leaves developing brown, water-soaked lesions. In advanced stages all tissues may be penetrated by the fungus. This disease is especially severe under greenhouse conditions where begonias are propagated. Growers should be sure to start with disease-free cuttings, use a sterilised medium and keep the growing area free of any type of weak or decaying plant material that might serve as a food source for the fungus.

Leaf Spots (fungi - several): Brown spots appear on foliage and reduce plant vigor. Most leaf spotting fungi thrive under high moisture conditions. Change the location of potted plants if conditions are overly wet and use appropriate protectant fungicides.

Powdery Mildew (fungus - Erysiphe cichoracearum): Affected leaves have a white powdery substance on the upper surfaces. This problem is occasional in nature but may occur when environmental conditions are ideal for disease development.

Bacterial Spot (bacterium - Xanthomonas campestris, pv. begoniae): Small blister-like spots appear on leaves. These become clear with age and may run together to form larger spots. Affected leaves may shed prematurely. Chemical control may be only partially effective.

Root Knot (nematode - Meloidogyne spp.): This nematode causes knots to form on roots. This problem can be prevented by using nematode-free planting stock and a sterilised potting medium.

Rust (fungus - Coleosporium solidaginis): Orange-red pustules form on the underside of leaves. In severe cases, foliage turns yellow and dies. Alternate host is pine where it produces a blister rust on the needles. Use rust resistant varieties when available. Fungicides should be used at the first sign of disease and continued on a 7-14 day schedule.

Aster Yellows (mycoplasma): First symptoms are observed as a yellowing or chlorotic appearance along the veins of young leaves. As the chlorosis becomes severe, defoliation occurs. Affected plants do not wilt or die, but have a spindly type growth which detracts from the plants overall appearance. Yellows may attack only a portion of the plant. Secondary shoots are formed profusely on infected plants. The disease is spread by leafhoppers. Control is obtained by insect control and removal of diseased plants.

CACTI AND SUCCULENTS

Many genera and species

Cactus Anthracnose (fungus - Colletotrichum (Gleosporium) spp.): This disease affects several kinds of cacti, Cereus, Echinocactus, Mammillaria, and particularly Opuntia (prickly pear). Infection results in a rather moist light brown rot which shows many light pink pustules on the surface. Spots are small at first, later enlarge and become covered by the small spore-producing pustules. Large areas may be affected, sometimes destroying entire plants. No satisfactory control is available, other than removing and destroying diseased cladodes as soon as noticed. In the greenhouse, soil from infected plants should be removed and benches disinfected. Spraying with a copper fungicide may help in checking the disease.

Charcoal Spot (fungus - Stevensea (Diplotheca) wrightii): Charcoal spot is a common and destructive disease of

Opuntia in Texas. Small spots, usually one-fourth inches or more in diameter, appear first. The spots are surrounded by a ring of small raised dots that are the fruiting structures. Spots later enlarge, but remain separated. There is no control for infected plants. Remove and destroy diseased specimens.

Dry Rot (fungi - Phyllosticta concava and Mycosphaerella spp.): Small black circular spots develop first, which later increase in size until they reach a diameter of one or two inches. Further advance is checked by the development of callus tissue. Minute fruiting structures are seen in the infected tissue. The disease is in part physiological, influenced chiefly by soil moisture. Remove and destroy diseased specimens.

Scorch or Sunscald (fungus - Hendersonia opuntiae): This disease is common and serious on prickly pear cactus (Opuntia). Spots at first are distinctly zoned, later enlarging until entire cladodes turn a reddish-brown and finally die. The center of the disease area is grayish-brown and cracked. Other fungi may also be present in the diseased area. No practical control has been developed.

Cotton Root Rot (fungus - Phymatotrichum omnivorum): Several members of the cactus family are susceptible to attack by the cotton root rot fungus. Infected plants die. When pulled from the soil the brown strands of the fungus can be found growing on the root surface.

Soft Rot (bacterium - Erwinia carotovora): The bacterium enters tissue through natural openings and wounds. Under conditions of high humidity, the bacteria reproduce quickly, spreading to healthy parts of the plant. Diseased tissue is watery, soft, black and deteriorates rapidly. If environmental conditions turn dry, the development of the disease may be checked. The best control is to avoid wounds, treat broken surfaces right away with a copper fungicide and avoid having plants in places where humidity is high.

Nematodes (Meloidogyne spp.): Most of the cacti and succulents are susceptible to infection by root knot nematodes. Infected roots show small galls which are typical of the disease and serve to identify it when clean, washed roots are observed.

Other Diseases (fungi): Other fungi known to cause disease on cacti are Fusarium oxysporum (Fusarium rot), *Macrophomina phaseolina* (Charcoal rot), *Septoria* spp., *Helminthosporium cactivorum*, and *Aspergillus alliaceus* (Stem and branch rot).

Scab (physiological): Particularly common on prickly pear cactus. Rusty colored, corky areas appear on the stems. Scab is thought to be a form of edema, resulting from overwatering and poor ventilation. Increase light and decrease humidity for control.

Stem Rot of Cacti (fungus - Drechslera cactivorum): Basal or top rot of seedling cacti that turns cactus into a shrunken mummy covered with brown spores. First symptoms are yellow spots. It can completely rot a plant in four days. The fungicide Captan should give some control.

CANDELABRA CACTUS

Euphorbia lactea

Stem Rot (fungus - Coniothyrium euphorbiae): A disease of *Euphorbia lactea,* especially when grown in greenhouses. The disease begins on the upper end of the branch and moves downward. Rotted areas are dark colored and soft. Once a branch is attacked it usually dies. The disease resembles a bacterial decay, but no foul odor is present. Black fruiting bodies of the fungus can be seen in the infected areas. The fungus becomes established in greenhouses and is spread by splashing water. Damaged or wounded plant are quickly infected. Control by eliminating diseased plants and water without splashing.

CANNA

Canna spp.

Rhizome Rot (fungus - Sclerotium rolfsii and Fusarium spp.): These organisms cause rhizomes to decay and stems to rot at the base. A cottony fungal growth may be present in rhizome rot.

Bacterial Bud Rot (bacterium - Xanthomonas cannae): Young shoots and leaves may be killed. Control of these organisms is through selection of healthy rootstocks. Damaged roots should be discarded. Plantings should be thinned periodically and replanted to increase vigor.

Mosaic (virus): Irregular light- and dark-green areas in leaves. Plants may be stunted and late in flowering. Destroy infected plants and control aphids. The variety - President is immune.

Other Diseases: Include Fungal Leaf Spot (*Alternaria* spp.) and Root Knot Nematodes (*Meloidogyne* spp.).

CARNATION

Dianthus spp.

Fusarium Wilt (fungus - Fusarium oxysporum f. spp. dianthi): First symptoms are a slow wilting of shoots, often on only one side of the plant. Wilting is accompanied by a discoloration of leaves, at first a light gray-green and finally a pale yellow color. When stems of diseased plants are split, a brown discoloration or streaks are evident in the vascular tissues. Symptoms develop rapidly and are more severe during periods of high temperatures. Most effective methods of control are propagation of disease-free cuttings and complete sterilisation of potting soil, tools and benches.

Rhizoctonia Stem Rot (fungus - Rhizoctonia solani): Foliage gradually becomes pale and wilted. A wet rot of the bark occurs at the soil line. Rot in the cortex is dry and corky with formations of sclerotia in the center of the stem. Soil sterilisation and soil drenches with fungicides are effective methods of control.

Leaf Spot (fungus - Septoria dianthi): Small circular light-brown spots with purplish brown border. Occurs on leaves and stems especially on lower portions of the plant. Small black specks (fruiting bodies) develop in the center of the spots. Tips of the leaves may die. Regular applications of a foliar fungicide spray in the greenhouse will control the disease.

Storage Rot or Botrytis Blight (fungus - Botrytis cinerea): Primarily a disease of cut flowers. Petals of infected flowers turn brown. May also cause a watersoaked flecking of the outer petals. Will also cause a soft rot of stem ends. Disease usually prevalent under high humidity. Spray or dips will effectively control this disease.

Streak (bacterium - Erwinia spp.): This bacterial disease usually starts at flowering time causing purplish-brown streaks on the stems, beginning from the base and extending upwards. The petioles, flowers, flower stalks, and pods are also streaked. Small round spots appear on the leaves running together gradually until the whole leaf is affected. Leaves may become dark brown or be entirely destroyed. Since this bacteria may be carried on the seed, chemical seed treatments and clean seed are advisable.

Other Bacterial Diseases: This host is also subject to crown gall caused by (bacterium - *Agrobacterium tumefaciens*) and leaf spot caused by (bacterium - *Pseudomonas syringae* pv. *pisi*).

Mosaic (Pea Mosaic Virus): Leaves are mottled, curled and yellowish. Flower stalks are short and flowers are streaked with white. Remove and burn infected plants. Control aphids which spread the disease.

Spotted Wilt (Tomato Spotted Wilt Virus - TSWV): Purplish spots on the stems and leaves develop after the leaves are mottled. Discolored or bleached spots may appear on the blossoms. Since the disease is carried by thrips, control the thrips and destroy all infected plants. Bud Drop Disease: This disease is due entirely to faulty culture; it results in the production of soft succulent growth and a lowered carbohydrate accumulation. A deficiency of phosphorus and potassium in the soil, coupled with low light intensity, are believed to aggravate the condition. If possible, provide extra light in the greenhouse during cloudy weather. Avoid overwatering and overfeeding with high nitrogenous fertilisers.

CHRYSANTHEMUM

Chrysanthemum morifolium

Fusarium Wilt (fungus - Fusarium spp.): When this disease occurs lower leaves turn yellow and wilt, followed by overall plant wilting and death of entire plant. Lower portion of stem near soil line turns dark brown-to-black with brown discoloration in vascular tissues. Planting disease-free cuttings, sterilisation of potting soil and benches offer best means of control.

Verticillium Wilt (fungus - Verticillium spp.): This disease is mainly in western areas of Texas. General wilting. Leaves become paler green, turn brown and die. Leaves remain attached to stem. In some case, wilting may not occur. First symptoms are a chlorosis of lower leaves which later turn brown. Infected plants are stunted with flowers reduced in size. Control same as for Fusarium Wilt.

Collar or Stem Rot (fungi - Pythium spp., Rhizoctonia spp., Fusarium spp.): Stem turns dark at soil line, entire stem is girdled and plant turns yellow and dies. Control measures include use of sterilised soil, disease-free cuttings, proper drainage and use of a soil drench.

Powdery Mildew (fungus - Erysiphe cichoracearum): Foliage covered with a whitish, ash-gray powdery growth.

With severe infectious leaves may become distorted and stunted. Regular foliar fungicide applications will control this disease.

Leaf Spots (fungi - Septoria spp., Alternaria spp., Cercospora spp., and others): Small yellow spots that gradually enlarge increasing to one inch in diameter. Center of spots turn brown. Infection takes place on lower leaves first. Regular applications of a foliar fungicide for control.

Leaf or Foliar Nematode (nematode - Alphelenchoides ritzemabosi): Yellowish-brown spotting of the leaves. Spots are bounded by larger veins. Spots coalesce until entire leaf is affected. Leaves die and fall. Symptoms start at base and progress up plant. Avoid wet humid conditions, do not water from top. Use a peat mulch. Use disease-free propagating material. Phosphate insecticides will control nematode on plant.

DAHLIA

Dahlia variabilis

Powdery Mildew (fungus - Erysiphe cichoracearum): Whitish-to-grayish growth covering upper surface of leaves. Avoid shaded locations and crowding of plants.

Ring Spot and Mosaic (viruses): Leaves are light green with mottled areas. Veins of leaf may appear light. Plants may be dwarfed and bushy. Leaves may have light green ring spots. Remove and destroy all infected plants as soon as detected. Tubers from virus-infected plants should never be used for planting purposes. Control aphids, leafhoppers and thrips.

Other Disease: Dahlias are susceptible to Verticillium Wilt, Bacterial Wilt, Root Rots, Root Knot nematodes Charcoal Rot and Southern Blight. In every case, diseased plants should not be used for propagation of new plants.

Tuber Rot (fungus - Botrytis cinerea): The most popular method of propagation is root division. To avoid root decay, dust fresh cuts with sulfur.

Flower Blight (fungus - Botrytis spp.): In moist weather the buds may be attacked by a soft rot and all infected parts may become covered with the grayish, powdery mold.

Dahlia Mosaic (virus): Infected plants are stunted, yellow and pale green bands will develop along the veins of affected leaves. The leaves may be smaller than normal, wrinkled or blistered.

Dahlia Ringspot (virus): Yellow circles, dime-sized, will develop on the leaves. As time passes these may merge into larger yellow to pale green areas and the centers of the rings may turn brown. This disease is carried in roots and cuttings from diseased plants. Dahlia mosaic and ringspot can also be transmitted by insects.

GERANIUM

Pelargonium hortorum

Cutting Rot (fungi and bacteria): Geraniums are grown from seed but often propagation is done with cuttings. Cuttings are susceptible to invasion by numerous soil-borne organisms, and, thus, treatment of cuttings with a fungicide is often necessary. Allowing the cuttings to "heal" before planting will help reduce stem rot. Wound healing takes place if cuttings are laid on damp sand in heavy shade for approximately three hours. Bacteria (*Xanthomonas* sp.) and fungi (*Pythium* sp., *Rhizoctonia* sp., *Fusarium* sp., and *Botrytis* sp.) cause stem rot. They may infect stems singularly or in combination.

Black Leg (fungus - Pythium spp.): Generally a disease of cuttings and young plants. Stems and petioles blacken and a soft rot develops. Rotting starts at the base of the stem and may extend well above the soil line. Plants wilt and die.

Symptoms progress rapidly. Control may require sterilizing potting mix and tools. Treat cuttings with a fungicide. A soil drench may retard spread of the disease in propagating benches.

Botrytis Blight (fungus - Botrytis cinerea): Affects blossoms, leaves and stems. On flowers, petals darken at edges and wilt prematurely. Affects central florets first. If humidity is high, spore masses may be found on flowers and leaves. Spots on leaves are irregular, brown and have a water-soaked appearance. Botrytis will cause a soft rot of cuttings. Proper sanitation in greenhouses by removing and destroying infected plant material will prevent spread. Use of a foliar fungicide when conditions are favorable for infection will help to prevent the disease. Improving ventilation and air circulation among the plants will also reduce infection.

Rust (fungus - Puccinia pelargonii): Rust occasionally is found on cultivated geraniums. Distinct, reddish pustules form on the underleaf surfaces in a circular pattern. The upper surface is yellow in areas where pustules form. Control rust by removing infected leaves and spraying with a fungicide.

Bacterial Leaf Spot and Stem Rot (bacterium - Xanthomonas campestris pv. pelargonii): Leaf spots begin as small, water-soaked spots on the underside of leaves. Spots become well defined and slightly sunken in a few days. Necrosis and wilting of the entire leaf follows, but the spots do not coalesce. Another symptom is rapid wilting of the leaf margin resulting in an angular pattern bound by the veins. Stem rot begins with one or two branches showing wilt. Eventually, the entire stem turns black with only a few leaves remaining at the terminals. Within the stem, the vascular fibers remain intact but the support tissue around the fibers is destroyed. Controls consist of using disease-free plants for cuttings, controlling the humidity in greenhouses, avoid wetting leaves when watering.

Viruses: Several viruses are known to affect geraniums. They cause mosaic patterns, mottling, crinkled or cupped leaves. Rogue out diseased plants as soon as they are noticed and use disease-free plants for propagation.

Oedema (physiological): At first, oedema appears as water-soaked spots on bottoms of leaves which later become corky and brown. The leaves may turn yellow and fall off. Oedema is caused by moist, warm soil and moist, cool air or cloudy conditions that result in more water being absorbed by roots than is being transpired by the leaves. As a result, cells swell and burst. Avoid over-watering and increase ventilation.

GLADIOLUS

Gladiolus hortulanus

Fusarium Rot (fungus - Fusarium oxysporum f. sp. gladioli): Corms decay with a brownish-to-black dry rot of the tissue. The foliage of affected plants first turn yellow and then brown. Roots are killed. When diseased corms are planted, many may rot before producing plants while other produce weak plants that soon die. Before planting, inspect and remove all corms that have discolored areas or lesions. Healthy appearing corm should be chemically treated before planting.

Scab (bacterium - Pseudomonas marginalis): Small definitely outlined lesions appear on diseased corms. The lesions are circular, water-soaked and pinpoint to one-fourth inch diameter. A gummy ooze is produced from these spots that is yellow-to-dark brown in color. The lesions are rather shallow. Badly affected plants may die while those less affected have an unthrifty appearance. Mites that occur in the soil seem to serve as a vector and should be controlled with soil treatment if this disease is a problem. Do not plant infected corms. Rotate areas where gladiolus are planted.

Botrytis Blight (fungus - Botrytis spp.): Leaves, stems, flowers and corms are infected. Leaf spots are variable in size and shape. Infected flowers decline rapidly and have grayish fungal growth on the surface. The tissues become moist and slimy. Under severe conditions the entire top may be killed. Infected corms have irregularly shaped sclerotia attached to them. Affected corms should be discarded. Use suggested fungicides for controlling the disease on above-ground plant parts.

Other Leaf Spots (Bacterium-Bacterial Leaf Blight - Xanthomonas spp.): Water-soaked spots appear on leaves with a bacterial ooze occurring during wet periods. (Fungus - *Septoria* Leaf Spot - *Septoria* gladioli) Small rot. (Fungus - *Stemphyllium* Leaf Blight - *Stemphyllium* spp.) Light green-to-yellow lesions form on leaves. Spots turn red.

Virus or Virus-Like Diseases (Virus-Mild Mosaic): Caused by the bean yellow mosaic virus. Produces angular light and dark green mottling of leaves and flower stems. Virus is transmitted by aphids. (Mycoplasma - Aster Yellow) Flower parts remain green and fail to form normal color pigment. Plants turn yellow and degenerate prematurely. (Virus - White Break Mosaic) Caused by the cucumber mottle virus. Causes flowers to be blotched, open slowly and fade early. Plants are stunted. (Virus - Ring Spot) Caused by the tobacco ring-spot virus. Ring spots appear on leaves but flower parts are not affected.

HOLLYHOCK

Althaea rosea

Anthracnose (fungus - Colletotrichum malvarum): This is the most destructive disease on Hollyhock. Fungus attacks leaves, stems and petioles. Typical fungal leaf spots, black blotches on petioles and stems. May also infect lower portions of stems and roots. High temperature and humidity favor development of the disease. Remove and destroy infected plant material and use a regular fungicide application.

Rust (fungus - Puccinia lobata): This is the most prevalent disease on Hollyhock in Texas and all portions of the plant are subject to infection. Small brown spots appear on underside of leaves. On upper leaf surface spots are yellow-to-orange and larger. Spots may coalesce and kill large portions on the leaf. Removal and destruction of infected plant material and regular applications of a fungicide will control rust.

Leaf Spots (fungi - Cercospora spp., Alternaria spp., Septoria spp.): Small angular grayish spots are scattered irregularly over leaves. Dead tissue may fall out leaving a shot hole appearance. Removal and destruction of infected plant material and regular applications of a foliar fungicide.

HYDRANGEA

Hydrangea spp.

Leaf Spots (fungi - Cercospora spp., Phyllosticta hydrangeae and others): Circular-to-irregular shaped spots are produced on leaves. Use foliar protectant fungicides as needed.

Powdery Mildew (fungus - Erysiphe polygoni): A white, powdery substance appears on leaf surfaces. Yellow spots that later turn brown may appear on the upper surface. If allowed to develop, it may get into newly developing buds and cause stunting. Use a foliar fungicide that has the ability to control powdery mildew.

Botrytis Blight (fungus - Botrytis cinerea): Botrytis is probably the most serious fungus attacking this plant because it causes severe problems in the flower buds. Infected flower parts may drop on leaves and cause infection of those structures. Cool, damp conditions with an abundance of fungal spores encourage the condition. This problem is especially severe at times in commercial greenhouses. Dispose of any weak or decaying plant tissue and use appropriate protectant fungicides if necessary.

Ring Spot (virus - Hydrangea ring-spot virus): Pale colored rings may appear on leaves and some varieties may have distorted leaves. Dispose of diseased plants. Commercial plant producers should use tolerant varieties.

Nematodes (nematodes - Meliodogyne spp., Pratylenchus spp., and Ditylenchus spp.): Nematode damage occurs when this plant is set in infested beds. Plants grown in pots would not be affected if nematode-free propagating stock and sterilised soil is used.

IRIS

Iris spp.

Leaf Spot (fungus - Cladosporium iridis): Irregular, eye-spots which vary in size and run together. They have a grayish center with dark brown borders. The disease usually is more or less confined to the upper surface of the leaf. Carefully remove and burn old iris leaves in the fall or early spring.

Leaf Spot (fungus - Heterosporium gracile): Common leafspot of iris in Central Texas. Disease shows up on both upper and lower leaf surfaces as 1/8 to 3/8 inch diameter, irregular spots with dark brown centers and outer edges. Remainder of spot is light tan. Good sanitation is best control. Removal and burning of old leaves each fall is desirable.

Basal Rot (fungus - Fusarium oxysporum): This soil fungus enters the roots of bulbous iris and then penetrates the base of the plant and the scales of the bulb. Bulbs should be dried rapidly to avoid possible injury. Cull and discard damaged bulbs.

Soft Rot (bacterium - Erwinia carotovora): A foul-smelling soft rot of the rhizome. At first, the leaves begin to turn brown and dieback progresses until the entire plant is dead. The disease is favored by wet, heavy soils or shady locations that tend to keep the ground continuously moist. Plant iris in a well-drained soil and avoid crowded conditions in the bed. In the fall, destroy all dead leaves and plant residue in the iris bed. In season when soft rot is severe, take up the rhizomes in August or September. Cut out the destroy all rotted portions of the rhizomes. Plant in a new location or sterilise the soil. Avoid excessive watering during the growing season. The disease is easily spread to new iris plantings during lifting and dividing operations by spades and other tools. Control iris insects such as the iris borer.

Rust (fungus - Puccinia spp.): Rusty-red, powdery pustules break out on both sides of the leaves, causing considerable damage to many varieties. Destroy all fallen leaves before growth starts in the spring.

Mosaic (virus): Yellow striping or mottling of the leaves and flowers. The potato and peach aphids transmit this virus. Discard all infected plants. Control aphids that transmit the virus.

Other Disease: Iris are attacked by several species of nematodes and *Sclerotium* rolpii.

PEONY

Paeoniae **spp.**

Botrytis Blight (fungus - Botrytis paeoniae): Most common disease of peony. Young shoots in all stages of growth, including the buds may suddenly wilt and fall over. Upon examination, a brown or blackish rot is seen at the base of the stem. This discoloration may extend down through the roots. Small buds attacked by the fungus turn black and dry up. Larger buds turn brown and may be covered with a brown spore mass. Large irregular dark brown lesions may occur on the leaves. Cool, rainy weather favors disease development. Splashing rain and insects spread the disease. Regular applications of a foliar fungicide, a drench, and destruction of infected plant material are recommended control measures.

Root and Stem Rot (fungus - Phytophthora cactorum): Infected parts are dark brown to black and leathery. Cankers appear along the stems and may cause them to fall over. May also cause a watery crown rot often destroying the entire plant. Roguing and destroying all infected plants is the only means of control.

Wilt (fungus - Verticillium albo-atrum): Plants gradually wilt and die during the blooming season. Brown discoloration of the water conducting tissues may be seen in cross sections of the roots or stems. Obtain disease-free plants, remove and destroy infected plants. Potting soil should be sterilised.

Mosaic (virus): Circular areas consisting of concentric bands of alternating dark and light green on leaves. Small necrotic spots may also form. Plants are not dwarfed. No control other than destroying infected plants.

Leaf Spot (fungus - Alternaria spp.): Leaves may have brown purplish or reddish irregular shaped spots. Leaves may yellow, wither and fall early.

Powdery Mildew (fungus - Erysiphe polygoni): Powdery, white mold on leaves. Leaves may be deformed, turn yellow and drop early. Collect and destroy fallen leaves and provide adequate spacing of plants.

Ring Spot (virus): Plants may be severely stunted with lemon-yellow to orange-amber spots, blotches or zoned rings. Young leaves may be distorted and plants may not flower. Do not propagate from infected plants.

PERIWINKLE DISEASES

Vinca rosea

Pythium Root and Stem Rot (fungus - Pythium spp.): Lesions produced on stems and roots with rapid progression of symptoms. Infected plants die soon after infection. Avoid overwatering and use suggested fungicide.

Other Root and Stem Rots (fungi - Rhizoctonia solani and others): Lesions occur on roots and lower stem. Periwinkle is more resistant to this problem than most annuals, but disease can occur when fungi are present in high populations. Rotate plantings and use recommended soil fungicides if needed.

Aster Yellows (mycoplasma): Infected plants are chlorotic and stunted. Control leafhoppers that transmit the disease-causing entity and rogue infected plants.

Canker and Dieback (fungus - Phomopsis sp.): Shoot tips will turn dark brown-to-black and die back to the soil. Favored by rainy weather. Preventative spray of copper fungicide starting in spring on new growth will control the fungus.

Phytophthora Aerial Blight (fungus - Phytophthora parasitica): Infection is generally characterised by leaf blight and upper stem deterioration. Infection is distinctly an aerial problem, not involving lower stem and root tissue. In most cases, leaves are shriveled and dull-gray in color. An aerial sunken stem canker can also develop. Symptoms develop rapidly in outdoor planted beds, especially during conditions of prolonged, rainy weather. Bark mulches or other mulch materials might help minimise splash dispersal of infested soil particles. No fungicides are currently cleared for use in control.

PETUNIA

Petunia hybrida

Damping-Off and Stem Rot (fungi - Rhizoctonia solani, Sclerotinia spp., Fusarium spp.): A damping-off or decay of plant at or below the soil line. A dark brown-to-black lesion on stems and crown. Use sterilised potting mix. A fungicide drench will help prevent spread.

Mosaic (virus - TMV): Crinkled leaves with yellow and green areas. Plant may be stunted. Destroy diseased plants. No control.

Crown Rot (fungus - Phytophthora parasitica): A black discoloration and dry rot of the crown and lateral branches occurs near the soil line. This is followed by wilting and death of the affected plants.

Aster Yellows (mycoplasma): Plants are yellowish, bushy, and dwarfed with many rosette-like, secondary shoots. Flowers often malformed and greenish. Leaves may curl, turn yellow or purple.

PHLOX

***Phlox* spp.**

Leaf Spots (fungi - Ascochyta spp., Cercospora spp., Macrophoma spp., Phyllosticta spp., Ramularia spp., Septoria spp., and Stemphylium spp.): Septoria divaricata is among the most common and most destructive. It attacks primarily the lower leaves, producing dark brown, circular spots up to one-fourth inch in diameter, the centers of which are light gray, almost white. Infected leaves dry up and die prematurely. For control, spray periodically with recommended fungicides and burn infected leaves.

Powdery Mildew (fungi - Erysiphe cichoracearum and Sphaerotheca humuli): White or grayish moldy growth on upper side of older leaves. Leaves gradually die, beginning on the lower part of the plant. For control, spray with recommended fungicides as soon as the mildew appears.

Rust (fungi - Puccinia douglasii and Uromyces acuminatus f. spp. polemonii): Neither rust is important enough to warrant control. Both rusts appear on leaves as raised blisters or pustules covered with a brown coating that ruptures releasing the brown spores.

Crown Rot (fungi - Sclerotium spp., Rhizoctonia spp. and Thielaviopsis spp.): These fungi are destructive on seedlings. They cause a rotting of the plant at the soil line. Sterilising potting soil is the best control.

Aster Yellows (mycoplasma): Flower petals often show white steaks and vein banding. Clusters of flowers or individual petals may turn into green, leafy structures.

Stem Blight (fungus - Pyrenochaeta spp.): Base of stem and roots may rot. They may be covered with a cottony mold. Avoid overwatering, overcrowding and planting in heavy, poorly drained soil.

Physiological Diseases: No organisms are associated with this condition. The disease is characterised by death of the lower leaves from the base progressively upward until the entire shoot is killed. The disease is most severe on old clumps and is entirely absent on seedlings or newly rooted cuttings. The trouble seems to be caused by excessive transpiration.

SNAPDRAGON

Antirrhinum majus

Anthracnose (fungus - Colletotrichum antirrhini): Large, light colored spots occur on leaves and stems. Spots frequently have black or brownish dots near centers. Entire leaves may be killed and often the whole plant dies from stem girdling. Spore bearing pustules appear as minute black pimples in the spots. Under a microscope, dark brown, hair-like outgrowths are visible on these pustules. Pull and burn affected plants as soon as disease appears. Spray with recommended fungicides to prevent spread. Water by irrigating surface of soil rather than wetting the foliage. Take cuttings only from healthy plants.

Rust (fungus - Puccinia antirrhini): Brown, powdery pustules appear on lower surface of leaves. Plants may be stunted and even killed. Propagate from healthy plants or grow from seed. Control insects that carry the fungus spores from one plant to another. Plants should be spaced so that they have plenty of air. Avoid wetting the foliage when watering. Keep greenhouse temperature above 70 degrees F for several days and not below 60 degrees F at night. Where foliage is wet when watering, spray at weekly intervals with recommended fungicide. Good coverage is necessary for best results.

Blight (fungus - Phyllosticta antirrhini): Cream colored or light brown circular spots appear on the leaves, sometimes surrounded by circles of other colors. The small black fruiting bodies of the fungus develop on the upper surface of the leaves at the center of the spots. Ash-gray spots covered with the black pycnidia develop on stems. These spots enlarge rapidly and turn dark brown. If the spots entirely encircle the stem, the plants wilt and frequently die. All parts of the plants in diseased beds should be removed and burned at the end of the season. Spray with recommended fungicides. Keep leaves dry and space more widely.

Wilt (fungus - Verticillium albo-atrum): Infected plants wilt slowly. At first, only certain branches show wilting. Plant in disease-free soil or sterilise soil with steam. When only a few plants appear infected, remove these together with surrounding soil. Water as little as possible to obtain good growth. Cotton burs used for mulch in West Texas may carry the fungus and should be well composted.

Gray Mold (fungus - Botrytis cinerea): This fungus causes a wilting of the flower stalks. A development of light brown areas on the stems at the base of the flower clusters. Under moist conditions, the spores of the fungus may develop in light brown or gray masses. Avoid excess humidity. Cut off and burn infected flower-stalks as soon as they are detected. Keep greenhouse and garden plantings free from debris because the fungus can live as a decay fungus under such conditions and its spores may be blown to living plants. Spray with recommended fungicides after the plants have been cut back to force a second crop.

Powdery Mildew (fungus - Oidium spp.): Largely confined to greenhouse-grown snapdragons, this mildew forms a white, powdery coating on both leaf surfaces and young stems. Spray with recommended fungicide.

Leaf Spot (fungus - Cercospora antirrhini): Previously reported only from Guatemala, this leaf spot was found in Florida in 1958. Spots are circular, 0.5 to 5 mm in diameter, dingy gray to white. No control measures have been developed but fungicides that control *Cercospora* spp. on other crops might be effective.

SWEET PEA

Lathyrus odoratus

Anthracnose (fungus - Glomerella cingulata): Leaves, flowers and shoots are at first marked with whitish spots. These spots spread extensively so that the parts attacked finally wilt and die. Seed pods shrivel and lose their color. Heavy infection results in much loss of leaves. The wilting works downward from the tips of the younger shoots, which become whitish and sometimes break. With a hand lens, many small salmon-colored pustules bearing great quantities of spores can be seen on the diseased parts. Gather and burn all infected plant parts after the flower season. Choose only seed pods which are plump and sound in appearance for planting. Seed for planting should be selected from disease-free pods. An occasional application of fungicide during the growing season helps minimise disease.

Powdery Mildew (fungus - Microsphera alni): Whitish, powdery growth on upper side of leaf surfaces. Leaves shrivel and drop off. Dust with sulfur or spray with recommended fungicide.

Downy Mildew (fungus - Peronospora trifoliorum): Under moist conditions, the leaves become covered with a grayish moldy growth. Thick-walled, brown resting spores develop within the moldy tissue. Since this is not a serious disease of this host in America, control measures have not been developed.

Leaf Spots (fungi - Colletotrichum pisi, Isariopsis griseola, Phyllosticta orobella, Mycosphaerella pinodes): Infection is not normally severe and spotted leaves can be picked off and destroyed as soon as they appear.

White Mold (fungus - Ramularia deusta f. spp. odorati): This mold, which may cover both sides of the leaves, may be mistaken for powdery mildew. Faint, dull-colored, irregular, elongated spots, which may be somewhat sunken, appear. The spots on the leaf margins, at first water-soaked in appearance, later have a reddish-brown discoloration. Tufts of spore-bearing hyphae develop from the stomatal openings between the epidermal cells. Spray with recommended fungicide as soon as white mold is noticed.

Fasciation (bacterium - Corynebacterium facians): Dense witches' broom, distortions and fasciations develop at the base of the stem or from below the soil line. Since the pathogen is seed-borne, plant clean seed.

TULIP

Tulips gesneriana

Bulb, Crown and Root Rot (fungi - Penicillium spp., Sclerotium rolfsii, Pythium spp., Fusarium spp., Rhizopus stolonifer, Rhizoctonia solani, Aspergillus spp.): Tulips may not come up or leaves may be yellow or reddish, stunted, and die because the bulb is rotted. Root may be slimy, soft, dry, hard or powdery in texture. Color may be dark or light gray, black or brown with moldy growth on bulb or scales. Remove and destroy infected plants. Avoid bruising in handling bulbs. Plant clean disease-free bulbs. Water and fertilise properly.

Botrytis Blight (fungus - Botrytis cinerea): Leaves and petals are affected with a brown decay, which often destroys large areas. Buds are sometimes blighted, or the flower stalks may tip over from a spot on the side where the fungus has attacked. A brownish-gray, powdery mold is produced on the surface of affected parts. All affected parts should be gathered and burned as soon as possible. Badly affected bulbs should be destroyed. Tulips should not be planted the next year in soil where this disease has been abundant. Plant grown in the greenhouse should be kept well ventilated.

Virus Diseases (Tulip breaking, Mosaic): Flowers may show changes in color with breaking, streaking, greening or darkening. Leaves and flower stems are mottled or streaked with light green or yellow areas. Plants are stunted and have poor vigor. Viruses are spread by cutting flowers and by several species of aphids. Destroy diseased plants when found. Control insects and weeds.

VIOLET AND PANSY

Viola odorata and Viola tricolor

Leaf Spots (fungi - Alternaria violae, Cercospora violae): Chlorotic spots on lower leaves, centers later turn brown. Usually on lower leaves first. Regular sprays with a foliar fungicide will control the disease.

Anthracnose (fungus - Colletotrichum violae - tricoloris): Browning or blotching of leaves. Spots may be small at first, but later enlarge. Spots may coalesce. Sunken, elongated dark brown to black lesions may appear on stems. In severe cases entire plant may be killed. Regular foliar fungicide applications will control this disease.

Scab (fungus - Sphaceloma violae): Small, green water-soaked dots. These later enlarge to form circular scabby spots. They may be gray-to-white surrounded by a dark-green border. Lesions may extend along a leaf veins and cause the veins to become distorted. Wilting of the leaves may result from petiole infections. A regular spray schedule with a foliar fungicide

Damping-Off (fungi - Rhizoctonia spp., Pythium spp., and Fusarium spp.): Plants wilt and die suddenly. A dark sunken area may be seen on the stem at or near the soil line. Use sterilised potting soil. Remove and destroy diseased plants.

ZINNIA

Zinnia elegans

Bacteria Wilt (bacterium - Pseudomonas solanacearum): This bacterium has been shown to cause a wilt of zinnias in Florida.

Blight (fungus - Alternaria zinniae): Small, reddish-brown spots with grayish centers are the first symptoms of this widespread and highly troublesome disease. Dark brown cankers also occur on the stems. Flowers are spotted and at times completely blighted. Because the causal fungus may be carried with the seed, treat the seed before planting with hot water at 125 degrees F for 30 minutes. Then cool and dry. Older seeds are more easily injured by this treatment. Use recommended protectant fungicide.

Powdery Mildew (fungus - Erysiphe cichoracearum): Whitish powdery growth on leaves and stems. Leaves die from base of plant. Affected plant lose vigor and cease growth. Spray with recommended fungicide as soon as mildew appears.

Leaf Spot (fungus - Cercospora zinniae): Small, roundish spots with gray or whitish centers. Several spots may run together, killing the entire leaf. Spray with recommended fungicides two or three times during the season.

Viruses: Many virus diseases affect zinnias. Among the most prevalent are curly top, caused by curly top virus;

mosaic, caused by the cucumber mosaic virus; and spotted wilt, cause by the tomato spotted wilt virus. Remove and destroy infected plants. Spray with insecticides every few weeks to control sucking insects which spread the viruses.

Nematodes: Two species of nematodes infest this host: *Meloidogyne incognita,* which causes root knot; and *Aphelenchoides ritzema-bosi,* which causes angular spots on the leaves.

Other Diseases: Zinnias are subject to other diseases including a blossom blight caused by a fungus, *Choanephora* spp., stem canker by *Botrytis cinerea* and damping-off by *Rhizoctonia solani.*

REFERENCES

Alfieri S.A Jr., El-Gholi N.E., "Phytophthora leaf blight of flowering dogwood, Cornus florida L.", *Plant Pathology Circular Gainesville,* 1993.

Hibben C.R., "Anthracnose threatens the flowering dogwood", *Methods for diagnosing and controlling this new disease,* Arnoldia Boston, 1990.

Sherald J.L, Stidham T.M, Hadidian J.M, Hoeldtke J.E., "Progression of the dogwood anthracnose epidemic and the status of flowering dogwood in Catoctin Mountain Park", *Plant-Disease,* 1996.

8

Diseases of Ground Covers

Groundcover is a plant used for the purpose of growing over an area of ground to hide it or to protect it from erosion or drought. In an ecosystem, the ground cover is the layer of vegetation below the shrub layer.

Strictly speaking, the most widespread groundcover are grasses of various types. In gardening terms, however, the term groundcover refers to non-grass plants that are used in place of grasses.

Four general types of plants are commonly used as groundcovers:

— Vines, which are woody plants with slender, spreading stems

— Herbaceous plants, or non-woody plants

— Shrubs of low-growing, spreading species

— Moss of larger, coarser species

AJUGA

Ajuga reptans

Root Knot Nematode (nematode - Meloidogyne incognita): Ajuga is very susceptible to root knot nematode. Infected plantings will begin to thin-out in spots and plants may die. Roots of infected plants will contain small knots caused by nematode feeding.

Southern Blight (fungus - Sclerotium rolfsii): Southern blight, also called Crown Rot, is a soil-borne disease of ajuga that will be more serious in poorly drained soils. Plants suddenly turn yellow and die in patches during warm, humid weather. Bases of stems rot and turn brown or black. Frequently, a white mold develops in which small reddish-tan bodies (sclerotia) develop. These sclerotia survive the winter and other unfavorable conditions.

ASIAN JASMINE

***Jasminium* spp.**

Crown Gall (bacterium - Agrobacterium tumefaciens): Plants will lack vigor, be stunted and woody galls will form near the base of the plants or on the roots.

Other Diseases (fungi): Other fungi known to cause disease on Jasmine in Texas are Phoma (blossom blight), *Colletotrichum gloeosporoides* (leaf spot).

CREEPING EUONYMUS

Euonymus fortunei

Anthracnose (fungus - Gloeosporium spp.): Brown lesions on the leaves, in which tiny fungus fruiting bodies may be seen with a hand lens. Apply dithiocarbamate or copper fungicides.

Crown Gall (bacterial - Agrobacterium tumefaciens): Both the stems and roots may have good size galls. Destroy heavily infected plants. Plants will lack vigor and may die back. If only a few galls, prune out and destroy. Sterilise pruner with 70% alcohol or 10% household bleach.

Powdery Mildew (fungus - Oidium spp.): White powdery mold or gray "felt" patches on leaves. Leaves may yellow, curl, and drop early. Prune to thin out shrub. Use preventive fungicide sprays.

Other Diseases (fungi): Other fungi known to cause disease on euonymus in Texas are *Phymatotrichum omnivorum* (root rot), *Exosporium concentricum*(leaf spot), and *Phyllosticta euonymi* (leaf spot).

DICHONDRA

***Dichondra* spp.**

Root Rot (fungus - Alternaria spp.): Poor drainage and winter cold may cause Alternaria root rot.

Rust (fungus -Puccinia dichondri): This rust, under humid conditions, may develop rapidly.

ENGLISH IVY

Hedra helix

Root and Stem Rot (fungus - Rhizoctonia solani): The fungus may attack roots, stems and lower leaves. symptoms on lower leaves first appear on those leaves in contact with the soil. Affected leaves are characterised by well-defined water-soaked areas which enlarge rapidly during cool, moist conditions. At first, leaves are dark green, they then wilt and turn dark-brown. Stems and roots may also be attacked. Infected areas are at first water-soaked in appearance and later become sunken and reddish-brown. Best control is obtained by planting only in sterilised beds. Beds should be well-drained. Soil fungicides applied as a drench around infected plants may be effective.

Root Rot (fungus - Phymatotrichum omnivorum): Affected plants wilt and die suddenly. Leaves of infected plants turn brown and remain on the plant.

Anthracnose (fungus - Colletotrichum omnivorum): A common disease of English ivy. Infected leaves have large tan-to-dark brown spots with a darker border. The center of infected areas may have small black spots formed by pycnidia (fruiting bodies of the fungus). Spots range in size from one-eighth inch to one-half inch in diameter. As the spots age, the centers may dry up and fall out giving a shredded or shot hole appearance. The fungus may also infect stems causing dark, sunken lesions. In some instances, the stem may be completely girdled causing death of the distal portion. Infected leaves should be picked off and destroyed. Avoid overhead watering. Foliar fungicide application at regular intervals are effective.

Bacterial Leaf Spot (bacteria - Xanthomonas spp.): Circular spots which are, at first, light green water-soaked areas. As the spots enlarge, they become brown-to-dark brown with red margins. Eventually, the center of the spots dry and crack. Young twigs and petioles are also attacked. Infection on stems and petioles causes a blackening and shriveling of the tissue. Cankers form and may girdle the stem. Tips of growing portions of the plant turn black and this blackening may extend downward into the older wood. Infected plants may fail to grow normally and have pale yellow-green foliage. Remove and destroy all infected plant tissue. Avoid overhead watering, overwatering and high temperatures, if possible.

Leaf Spots (fungi - Ramularia hedericola, Macrophoma spp., Glomerella cingulata, Phyllosticta concentrica): Brown lesions will occur on the leaves and may also infect the stem, causing girdling, collapse and death of apical portion. Infected leaves can be picked off and burned. Remaining foliage should be sprayed with a fungicide.

Sooty Mold (fungus, Capnodium sp., and others): The sooty mold fungus may develop on English ivy growing beneath certain trees. This mold lives on secretions of aphids and scale insects which fall from the tree above. This mold seldom causes damage. Control insects to prevent this problem.

HONEYSUCKLE

***Lonicera* spp.**

Honeysuckle is a hardy plant that is not plagued with any major disease. Those diseases reported on honeysuckle rarely become severe enough to warrant control fungicides.

Leaf Spots (fungi - Cercospora spp., Septoria sambucina and others): Avoid overhead irrigation. If spraying is necessary, apply a recommended fungicide.

Leaf Blight (fungus - Glomerularia lonicerae): This fungus blights the leaves during rainy seasons with light-colored, circular spots that can cause leaf death. Control by spraying with recommended fungicides.

Powdery Mildew (fungi - Microsphaera alni and Erysiphe polygoni): These powdery mildew form a white, powdery coating on both leaf surfaces and young stems. Microsphaera is rather widespread in the United States while *Erysiphe* occurs primarily in the western United States. Control by spraying with recommended fungicides.

Cotton Root Rot (fungus - Phymatotrichum omnivorum): Honeysuckle is only moderately susceptible to cotton root rot.

Leaf Spots or Blights (fungus - several): Small-to-large round to irregular spots and blotches on leaves and pods. Leaves may be blighted and drop.

Rarely Observed Diseases: Honeysuckle is also susceptible to thread blight, caused by the fungus *Pellicularia koleroga,* twig blight caused by the fungus *Phoma mariae,* rust caused by the fungus *Puccinia festucae,* hairy root caused by the bacterium *Agrobacterium rhizogenes,* and crown gall caused by the bacterium *Agrobacterium tumefaciens.*

LIRIOPE

***Liriope* sp.**

Leaf Anthracnose (fungus - Colletotrichum sp.): Anthracnose is characterised by brown, necrotic lesions that usually appear along leaf margins and leaf tips. Foliar wetness can intensify infection and subsequent damage. Protective fungicides can help reduce damage if persistent, severe infection occurs.

MOSS PHLOX

Phlox xubulata

Leaf Spot (fungus - Ascochyta phlogis): Remove spotted leaves. spray healthy leaves with recommended fungicide.

Powdery Mildew (fungus): Whitish, powdery mold patches on upper side of leaves and stems from midsummer on. Leaves may shrivel and drop early. Control with recommended fungicides.

Crown Rot (fungi - Sclerotium rolfsii, Sclerotium sclerotirum and Thielaviopsis basicola): Serious losses to phlox seedlings occur from these soil-borne fungi. Start in pasteurised soil.

Root Rots (fungi - Thielaviopsis basicola and Phymatotrichum omnivorum): Not usually serious enough to warrant control.

Rusts (fungi - Puccinia douglasii and Uromyces acuminatus): Not usually serious enough to warrant control.

OXALIS

Oxalis spp.

Rust (fungus - Puccinia sorghi): The importance of this rust on oxalis is that its alternate host is corn. On oxalis, the rust first appears as neat, yellowish dots near the margins of the leaves; later pale orange pustules break out on the leaf. Usually does not cause enough damage to warrant control.

Powdery Mildew (fungus- Microsphaera russellii): White, powdery blotches on foliage.

VINCA

Vinca major, Vinca minor

Canker and Dieback The fungi causing this disease (Phoma sp. and Phomopsis sp.) cause the shoot tips to become dark brown, wilt, and die back to the surface of the soil. Some affected stems are black in color. Small, black, dot-like, fruiting bodies develop on the surface of the diseased stem tissue.

Wet weather or frequent overhead watering favors the disease. Cultural control is very important. For chemical control, some copper or mancozeb (FORE) containing fungicides are labeled and can be used according to labeled instructions. Spray affected plants once a month from bud break in the spring until plant growth slows in mid or late summer.

Leaf Spot (fungus - Alternaria spp.): Brown circular-to-oval spots on leaves. Control with copper or other recommended fungicides.

CONTROL MEASURES FOR INFECTIOUS GROUNDCOVER DISEASES

Infectious plant diseases are best managed via an integrated approach employing many strategies. Some of these strategies are aimed toward sanitation and removal of infected plant tissue or the pathogen. Other strategies are aimed at enhancing the vigor and natural resistance of the plants.

1. Always select and plant high quality plant material. Select plants that are growing well and are free of nutritional deficiencies. If any disease is noted in the plants, do not purchase them.
2. For established beds, remove infected plants and prune out diseased parts of plants. Discard or burn infected material.

3. Thin planting to remove dense growth. Thin in fall during dry weather.
4. Fertilise appropriately to maintain plant vigor. If excess fertilisation causes dense growth, plants may need to be thinned.
5. Remove dead leaves and other accumulated plant debris. Do this very early in the spring.
6. Water to prevent weakened, drought stressed plants. Water early in the day so that drying occurs before evening. Avoid splashing of water.
7. Use fungicides as already mentioned above. Spray to thoroughly wet the plants. Use a spreader-sticker spray additive if the fungicide you have selected is a powder. Many of the products mentioned above may not be available in small packages for home landscape use. If such is the case, home horticulturists should contract a professional plant care service to make the applications.

Choosing, Planting and Care for Groundcover

Ground covers are low-growing plants, usually less than 12 inches tall, that spread to form dense mats which bind and hold the soil in place. Many also have a season of effective bloom. Turfgrass is the most common ground cover. However, sites that are less suitable for turfgrass, such as slopes, steep banks, and shaded areas, can often grow other ground covers successfully.

A well-established ground cover generally requires less maintenance than the typical turfgrass lawn. Ground covers are not totally maintenance-free, nor will they always tolerate conditions that will not support a turf.

There are many ways of trying to control weeds. We can remove them by hoeing them out or by burning them with a flame gun. We can try to prevent them growing by putting down mulches. Each of these methods have their advantages and disadvantages and each has situations where they are applicable. In this section we will look at another method of controlling weeds - by growing carefully selected plants that can spread to cover the ground and that will out-compete the weeds

Its primary advantages include:

1. By covering the ground with a carpet of vegetation it prevents the germination of weed seeds.
2. When established, it prevents the ingress of perennial weeds.
3. It protects the soil from erosion and water loss. This can be particularly useful on steep slopes.
4. It acts as an insulating cover for the soil, keeping it cooler in the summer and warmer in the winter.
5. It is a living mulch that helps to build up humus levels in the soil.

Selecting a Ground Cover

When deciding which ground cover to plant, consider the plant's height, spread, blooming period, seasonal foliage color, texture, drought tolerance, and whether it is deciduous or evergreen. The planting site also will affect your choice because each ground cover has specific requirements.

Many ground covers develop crown rot in heavy clay soils that don't drain well. Adding coarse organic matter, such as composted or aged manure, peat moss, or compost can improve the drainage of heavy clay soils and increase the water-holding capacity of light sandy soils. Roto-till or spade the soil to an 8-inch depth and thoroughly incorporate the organic matter at a rate of two to three bushels per 100 square feet.

Soil preparation can cause serious erosion on steep slopes. Working the soil across the slopes in 12- to 24-inch wide bands alternated with undisturbed soil will hep reduce erosion. Mulching the slope with a fiber mat will further reduce erosion while conserving moisture and reducing weed competition.

Planting

Ground covers should be planted in the spring or fall. Containerised stock can be planted anytime during the growing season if proper planting and maintenance procedures are followed.

The number of plants needed depends on the spacing of the individual plants their rate of growth and how fast you want the space to fill in. You may want to densely plant a small, highly visible area for rapid fill, but a large area may require more sparse planting, with several years being allowed for covering the site.

Space between plants in inches using 50 plants.

4" between plants 7 sq. ft.

6" between plants 16 sq. ft.

8" between plants 28 sq. ft.

12" between plants 64 sq ft.

When planting rooted cuttings, don't allow them to dry out. Plant them at the proper depth and firm the soil around the roots to ensure good soil-root contact. Watering small sections as you plant will prevent the roots from drying out.

Ground covers can be planted in either straight or staggered rows. Staggered rows help reduce erosion on slopes by retarding run-off.

Watering

Water new plantings thoroughly and regularly until the roots become established. Once established, water as needed for the ground cover you've selected—each type has different requirements—but apply at least 1 inch of water at each watering to moisten the entire root zone. Applying less water results in a shallow root system which makes the planting more susceptible to drought injury.

Controlling Weeds

Control weeds to enable the newly planted ground cover to form a dense mat. When properly established, ground covers can successfully compete with weeds. Hand weeding is effective, but can be time consuming. Mulching is a practical way to reduce weeds. Nonaggressive annual flowers can be used to fill in spaces in a new planting. They help shade out weeds but do not inhibit the spreading growth of the ground cover. Be sure to provide enough moisture for both plant types so competition is not a problem.

Preemergent herbicides can be used to control weeds in large plantings. Dacthal, Treflan, and Dymid can be used on most ground covers, but always check the label for specific information on the particular plant. If perennial weeds, such as bindweed or annual grasses, are a problem, consider fumigating the site with Vapam after preparing the soil. Wait at least two weeks before planting and cultivate the soil again to be sure the chemical has dissipated.

Pruning

Pruning will stimulate new growth on most ground covers by causing buds to break from the base or along the plant's stem. At planting time, prune back the growth by one half (more on trailing plants such as ivy or periwinkle) to promote branching. Annual pruning will control ground covers and keep them attractive. Generally, the best time to prune is just after new growth has begun in the spring. A mower set very high, nylon cord trimmer, or hand clippers can be used. Rejuvenation and size control are important for plants such as wintercreeper, St. John's-wort and carpet bugle. Cutting back after flowering can induce repeat flowering in sun rose, alyssum and cottage pinks. Pruning also can help control diseases by removing infected foliage.

Mulching

Mulching serves many purposes in a ground cover planting. Two to four inches of mulch around the plants helps control weeds, maintains a more constant soil temperature, conserves soil moisture and reduces erosion on slopes. Compost, leaf mold, well-rotted manure, bark or wood chips, peat moss, and sawdust can be used as mulches. However, mulches that have not completely decomposed, such as fresh sawdust, will temporarily draw nitrogen from the soil. An application of additional nitrogen will make up for this excess use. Mulching helps to protect plants from winter injury resulting from uneven soil temperatures and soil heaving. Most evergreen ground covers need protection from winter desiccation (drying). Apply mulch loosely after the ground has frozen and remove it in the spring before growth begins. Weed-free straw, peat moss and pine tree boughs can be used for winter mulch.

REFERENCES

Hottes, Alfred C., *Climbers and Ground Covers*, A. T. De La Mare Company, Inc., New York,1947.

MacKenzie, David S., *Perennial Ground Covers*, Timber Press. Portland, Oregon, 1997.

Westcott, Cynthia and Horst, R. Kenneth, *Westcott's Plant Disease Handbook*, Van Norstrand Reinhold Company, New York, 1978.

9

Diseases of Shrubs

A large number of plants can be either shrubs or trees, depending on the growing conditions they experience. Small, low shrubs such as lavender, periwinkle and thyme are often termed subshrubs.

An area of cultivated shrubs in a park or garden is known as a shrubbery. When clipped as topiary, shrubs generally have dense foliage and many small leafy branches growing close together. Many shrubs respond well to renewal pruning, in which hard cutting back to a 'stool' results in long new stems known as "canes". Other shrubs respond better to selective pruning to reveal their structure and character.

Shrubs in common garden practice are generally broad-leaved plants, though some smaller conifers such as Mountain Pine and Common Juniper are also shrubby in structure. Shrubs can be either deciduous or evergreen.

ALTHEA (ROSE OF SHARON)

Hibiscus **spp.**

Flower Bud Drop (physiological): Open flowers and buds may drop suddenly when one moisture extreme follows another. For example, drought stress followed by adequate moisture may cause such drop. Other extremes in growing conditions may also contribute. Avoid drought stress.

Leaf Spot (fungi - Cercospora spp., Phyllosticta spp. and others): Circular spots occur on leaves which results in their shedding from the plant. Use appropriate foliar fungicides to prevent damage.

Root Knot Nematodes (Meloidogyne spp.): Knots or galls form on roots causing the plant to be inefficient in absorbing water and nutrients. Leaves are small and the plants fail to make normal growth. Use caution when planting to be sure that planting stock is not infested. If yard soil is contaminated, plant shrub in a large container of sterilised soil. This is probably the most serious problem experienced by althea.

Cotton Root Rot (fungus - Phymatotrichum omnivorum): Althea is very susceptible to attack by the cotton root rot fungus that occurs in the soils of central Texas. If this disease is a problem in the area, it may be necessary to grow the plant in a large container of sterilised potting mix or soil.

Leaf Rust (fungus - Kuehneola malvicola): Infection by the leaf rust fungus causes chlorosis and leaf spotting. Infection is characterised by yellow-orange pustules on the lower leaf surface. Fungicide control is usually not necessary.

AZALEA

***Rhododendron* spp.**

Anthracnose - Leafspot and Dieback (fungus - Glomerella cingulata): Symptoms on leaves are characterised by numerous small, round spots on both leaf surfaces. The spots are light to rusty-brown with a definite margin. Under severe conditions, the spots may coalesce and form large irregular spots. Premature defoliation may occur. Branches may suddenly wilt and die. Upon close examination, stem cankers that have girdled the stem may be observed. Pink pustules or spore masses may be seen in these cankers during moist periods. Plant in well drained soil and avoid excessive nitrogen fertilisation. When dieback occurs, prune out and destroy diseased twigs. Preventive sprays with a foliar fungicide are effective.

Petal Blight (fungus - Ovulinia azaleae): The first evidence of infection is the appearance of small, brown irregular-shaped spots on the petals. Spots are white on colored flowers and brown on white flowers. Affected petals have a veined appearance and remain firm. Flowers may have a freckled appearance. The flower collapses and sticks to the foliage. Small, black fruiting bodies (sclerotia) form on blighted flowers. Where flower blight has previously occurred, spray soil and litter around plants with a recommended fungicide four weeks before bloom. Spray at regular intervals during the blooming period.

Leaf Spot (fungi - Cercospora sp., Phyllosticta sp.): Spots of various sizes and color on leaves throughout the year. Fungicidal sprays at 7-10 day intervals during periods of high humidity will prevent serious foliage damage.

Leaf Gall (fungus - Exobasidium sp.): Distorted leaf and bud growth developing in April and may. Affected parts later become covered with a whitish mold. Spray plants with protective fungicide just before flower buds open and at 10 day intervals until flowering ceases. Remove and burn affected parts.

Root and Crown Rot (fungus - Phytophthora cryptogea): Affected plants may wilt suddenly and die or they may be unthrifty if only a portion of the root system is infected. A brown discoloration of the wood may be seen in the basal portions of the stems and in the main roots. Plants are much more susceptible when planted in poorly drained areas or planted too deeply. Plant only in well-drained and well-aerated soil. Plant only as deep as the plants originally grew. Drench with a soil fungicide before plants are seriously affected.

Physiological Disorders: Leaf margins and leaf tips turn brown. Plant are unthrifty or fail to grow. Principal cause is improper soil moisture relationship; i.e., soil allowed to become too dry or soil waterlogged to exclude oxygen. Contributing factors may be salt toxicity from over-fertilisation, "root-bound" plants unable to develop normal root systems, soil alkalinity due to leaching of lime from masonry of new residences or the use of alkaline water from certain deep well sources. Select healthy bushes with unrestricted root systems showing active growth for planting. Plant in well-drained soil avoiding sites exposed to full sunlight or to the west sun. Maintain high amount of organic matter in soil and adequate mulch to preserve uniform soil moisture in the root zone.

BOXWOOD

Buxus spp.

Decline, Dieback, Root Rots, Twig Blight, Cankers (fungi - Armillaria spp., Fusarium spp., Gandoderma spp., Nectria spp., Phymatotrichum spp., Rhizoctonia spp.): Infected branches often start growth later in spring than normal ones. Leaves on such branches usually curl upward close to the stem and turn light grayish-green or bronze and finally straw-colored. Leaves may wither and drop early leaving bare twigs. Roots may be decayed. The bark at base of branches may die and slough off. Before growth starts in spring, remove and destroy all leaves on ground and lodged in twig crotches. Prune out dead twigs and severely cankered branches as soon as noticed. Spray with a fungicide just after removing dead leaves and branches and as new leaves begin to develop.

Leaf Spots, Leaf Tip Blights, Leaf Cast (fungi - Diplodia spp., Phomaspp., Phyllosticta spp.): Leaves variously spotted may become pale or straw-colored, sometimes dull tan or brown, starting at margins and tips. Many leaves drop early. Young twigs may die. Begin spraying with a foliar fungicide as new leaves emerge.

Nematodes (Lesion, Ring, Stubby Root, Stunt and Root Knot): Plants are weakened, stunted and lack vigor, may wilt on hot days and recover at night. Leaves may be pale green to bronze or orange. Plants gradually decline and branches may die back. Roots will be stunted, often bushy, dark and may have knots or galls on them. Fumigate planting site before replanting boxwood in infected soil.

Root Knot (nematode - Meloidogyne spp.): Plants usually lacking vigor are often stunted and yellow. Small-to-large swellings, galls or knots develop on the roots. These are usually round and cannot be broken off like nodules on legume roots. This is a major limiting factor in growing boxwood.

Root Rot (fungus - Phytophthora cinnamomi): Foliage has an off-color followed by sudden wilting and death of the entire plant. Usually infected plants cannot be saved.

Winter Injury (non-pathological): Leaves may turn bronze colored, rusty brown or red with dead areas around margins. Leaves will usually begin to dry in the late spring. Leaves, twigs and even entire plants may die back. Plants that are not watered during the wintertime are more susceptible to winter injury. Using mulches around the plants will help prevent freezing.

CAMELLIA

Camellia japonica

Canker and Dieback (fungus - Glomerella cingulata): Sudden wilting of branches is usually the first indication of disease. Gray blotches appear on bark of stem or branches. Underlying wood dies and bark may split to form open wounds or cankers. Pinkish pustules or spore masses may be seen in these cankers during moist periods. Leaves on affected branches turn chlorotic and branch tips die back. Plant in well-drained soil that has the proper pH and fertility level. Avoid excessive nitrogen fertilisation. When dieback occurs, prune out and destroy diseased twigs. Prune several inches below visible canker, and use a commercial wound dressing on all cuts. Spraying plants during wet periods and during the period of normal leaf drop with a fungicide will help control the disease.

Petal Blight (fungus - Sclerotinia camelliae): Petal blight affects only the flower portion of the plant. The first evidence of infection is the appearance of small, brown, irregular-shaped spots on the petals. Affected petals have a "veined" appearance and the tissue remains firm. Entire flower turns brown in 24-48 hours and usually drops. The disease generally occurs during the spring. Remove and destroy all infected plant tissue. Apply fungicide to the soil just prior to bloom to prevent germination of overwintering spores. Thoroughly cover all areas under the plant and at least a 10 foot radius around each plant. Applications of a foliar fungicide at the first sign of disease and at 7-14 day intervals is effective.

Leaf Gall (fungus - Exobasidium camelliae): Young leaves and buds may be infected. Leaves become thickened and succulent and may be larger than normal. The color changes from light green to nearly white. Little damage occurs but the plant appears abnormal or diseased. Pick young galls and destroy. Spray with fungicide before leaves open.

Leaf Spots (fungi - Phyllosticta camelliae, Pestalotia guepini and algae -Cephaleuros virescens): Various sized spots on the leaves. Small, black pinpoint dots may be in the dead leaf areas. Control requires preventive fungicidal sprays at two week intervals. Include a spreader sticker to increase effectiveness of sprays.

Ring Spots (virus): Occurs mostly in East Texas. Light green or yellowish rings surround islands of green in the leaves. Destroy diseased plants. Propagate from disease-free plants.

Yellow Mottle Leaf (virus): Irregular, yellow patterns on some leaves. No cure available. Propagate from disease-free plants.

Bud Drop (physiological): May occur as a result of overwatering, inadequate moisture in the soil, insufficient light, excessively high temperatures, severe freezing during the winter or a pot-bound condition of the roots. Also, certain camellia varieties shed their unopened bloom buds as a result of insufficient cold during the winter. Fertilise and water properly. Keep plants healthy by controlling diseases and insects.

Oedema or Scab (physiological): Corky bumps or raised areas on leaves. Usually due to excessive water in soil.

Sunburn (physiological): Leaves have brown or faded green areas. Occurs on side of bush directly hit by the sun. Particularly a problem on bushes transplanted from a shaded to a sunny location.

COTONEASTER

***Cotoneaster* spp.**

Fire Blight (bacterium -Erwinia amylovora): New shoots suddenly appearas if scorched by fire. Brown or blackened leaves cling to twigs. Slightly sunken, girdling, discolored cankers on twigs, branches and trunk. Prune and burn affected parts.

Leaf Spots (fungi - Entomosporium maculatum, Phyllosticta cotoneastri): With Phyllosticta, leaves will have reddish-brown spots bordered by darker zones, outside of which is a zone of dark red. There may also appear within the spots small, black fruiting structures of the fungus.

Canker (fungus - Physalospora obtusa): This fungus causes cankers and twig blight on cotoneaster. Prune and destroy infected branches.

CRAPEMYRTLE

Lagerstroemia indicia

Powdery Mildew (fungus - Erysiphe lagerstroemiae): Powdery mildew is very common on crapemyrtle. It is particularly active in the spring and fall months. White-to-grayish moldy growth develops on leaves and new shoots. Dust or spray with a recommended fungicide at first appearance of mildew. Dallas red is an old variety with good mildew resistance. Check on availability of newer varieties with mildew resistance.

Leaf Spots (fungi - Cercospora spp., Phyllosticta spp.): Leaf spot fungi seldom cause severe damage. If needed, apply a recommended fungicide.

ELAEAGNUS

Elaeagnus spp.

Wet Feet (physiological): Elaeagnus (Russian olive) is a hardy woody shrub with few disease problems. It is, however, vulnerable to excessively wet soils. Elaeagnus should be planted on well-drained sites and not in low spots or in soils that retain water.

Leaf Spots (fungi - Cercospora carii, Cercospora elaeagni, Phyllosticta argyrea; Septoria argyrea): Spots may vary in size, color and shape. These fungi are rarely severe enough to warrant control measures. Prune out heavily infected foliage if it becomes unsightly.

Southern Blight (fungus - Sclerotium rolfsii): Southern blight primarily attacks *elaeagnus* in the seedling stage. Plants usually wilt as the stems rot at or below the soil line. A white, cottony mold may be evident on the lower stems near the ground line. Brown fruiting bodies (sclerotia) are often scattered through the cottony fungus growing on the stem. This disease does not normally attack older plants.

Cotton Root Rot (fungus -Phymatotrichum omnivorum): As with most woody ornamentals, elaeagnus will die quite rapidly when infected with the cotton root rot fungus. Plants will die so rapidly that leaves are not shed but are retained on the plant. Cutting the base of the plant with a knife soon after it dies will show that the base of the plant and the roots died before the upper part of the plant. The woody portion of the crown will be darker than in the branches and stems. No controls are known.

EUONYMOUS

***Euonymous* spp.**

Powdery Mildew (fungi - Oidium euonymi-japonici and Microsphaera alni): Powdery mildew is the most common and possibly the most difficult disease to control on Euonymous. It forms a flat, whitish growth on the leaves which can be easily rubbed off. Leaves may yellow slightly and drop off, but heavy defoliation is not a characteristic symptom. The fungus only invades the epidermal cells and, thus, leaves normally stay attached long after they are infected. Preventing infection of the new growth in the spring and planting in a proper site are keys to controlling powdery mildew. Cultural controls include: (1) planting in a sunny area; (2) not crowding the plants; (3) avoiding overhead watering; (4) pruning out and destroying heavily disease branches; and (5) raking and destroying fallen leaves. Chemical applications should begin before new foliage is infected. Fungicides will protect foliage from infection but will not remove the dense, white fungal growth from infected leaves. Adding a wetting agent to the spray solution will greatly improve coverage. Liquid household detergents make good wetting agents when one teaspoon is added for each gallon of spray solution.

Crown Gall (bacterium - Agrobacterium tumefaciens): Large, rounded galls with an irregular rough surface, appear on roots or stems. Plants lack vigor and may die back. Avoid wounding stems or roots. Prune out and destroy affected plant parts. Sterilise pruning shears in 90% alcohol or 10% bleach after each cut. Dig up and destroy severely affected plants.

Anthracnose (fungus - Colletotrichum spp.): Symptoms consist of small, brownish spots on the leaves with light colored centers. Tiny cracks in the leaf spots indicate fruiting structures of the fungus. Overhead watering makes the disease worse and considerable defoliation can result. Fungicide applications will help control this disease.

Leaf Spots (fungi - Cercospora destructiva, Phyllosticta euonymi, Septoria euonymi): Various sizes and colored spots on leaves. Collect and burn fallen leaves. Prune to thin out shrubs. Protect with fungicide sprays when needed.

Cotton Root Rot (fungus - Phymatotrichum omnivorum): Plants suddenly wilt and die. when plants are removed from the soil, the bark on the roots is decayed.

Oedema (physiological): Small, raised masses of tissue on the underside of leaves. the raised mass of tissue later becomes rough and corky. May sometimes occur on stems. Control by reducing moisture level. In greenhouses, reduce humidity, improve air circulation and avoid overwatering during overcast periods.

GARDENIA (CAPE JASMINE)

Gardenia jasminoides

Canker (fungus - Phomopsis gardeniae): One of the most common gardenia diseases. The main stem is swollen near or below the soil line. The bark becomes corky and contains numerous longitudinal cracks in the cankered area. The stem above the canker is bright yellow in contrast to normal greenish white. If the humidity is high, a yellowish exudate may be seen on the surface. Affected plants are stunted and die slowly. Destroy all diseased plants to prevent spread of the disease. Place new plants in a different location. Disease is easily spread on propagating knives.

Bacterial Leaf Spot (bacteria - Pseudomonas gardeniae, Xanthomonas campestris cv. maculifolium-gardeniae): Small, rounded ovoid spots on young tender leaves. As the spots enlarge, the center is at first pale yellow and later become reddish-brown surrounded by a yellow halo. Margins of the lesions are thickened and water-soaked in appearance. Spots may coalesce to form large, irregular-shaped spots. Severe infection may cause defoliation. Avoid overhead watering. The disease is spread by taking cuttings from infected plants. Use sterilised soil and pots.

Rhizoctonia Leaf Spot (fungus - Rhizoctonia spp.): Leaves have tan to brown spots up to one-fourth inch in diameter. Spots are circular and zoned. The disease begins on the older leaves and spreads upward when the plants are watered excessively or when air circulates poorly because of overcrowding. Diseased leaves should be destroyed and sterilised soil should be used. Disease-free plants should be used for propagation. Avoid wetting foliage when watering.

Leaf Spot (fungi - Cercospora spp., Phyllosticta spp.): Spots of various sizes on leaves throughout the year. Spots may be small, dark-brown necrotic areas surrounded by a yellow halo. In severe cases, premature leaf drop may occur. control is obtained by spraying with a foliar fungicide at regular intervals.

Sooty Mold (fungus - Capnodium spp.): Black, thin layers of the fungus form over the upper surface of leaves. Sooty mold is caused by a fungus that grows on sugary exudate from white flies. Control white fly.

Bud Drop: Abnormal dropping of buds. Occurs during period of high night temperatures or during periods of low light intensity. Some bud drop is a natural condition. Every effort should be made to keep the soil uniformly moist, but not wet, during flowering.

Powdery Mildew (fungus - Erysiphe polygoni): White powdery spots on leaves. Use preventive fungicide.

HOLLY AND YAUPON

***Ilex* spp.**

Leaf Spots (fungi - Cercospora spp., Phyllosticta spp.): Brown or purple spots occur on leaves. Spray with recommended fungicides at 7-10 day intervals. Include a spreader-sticker or detergent in spray mix to help wet leaves thoroughly.

LIGUSTRUM

***Ligustrum* spp.**

Anthracnose (fungus - Glomerella cingulata): Leaves dry out and cling to the stem. Twigs are blighted and cankers are formed at the base of the main stem. cankers may be spotted with pinkish pustules. The bark and wood of diseased portions become brown and the bark on the cankers may split, exposing the wood, Death occurs when the cankers completely encircle the .twigs or stems. The following varieties are resistant: *Amur privet, L. amurense; Ibota privet, L. ibota; regal privet, L. ibota regelianum; and California privet, L. ovalifolium.* For climates in which winter killing is not a factor, the California privet is preferred for planting.

Green Scurf (alga - Cephaleürus virescens): Green scurf or algal leaf spot appears on twigs and limbs. Spray or dust with a copper fungicide recommended for general use on ornamental plants being careful to cover the underside of the leaves. Add a spreader-sticker to the spray.

Leaf Spots (fungi - Exosporium concentricum, Cercospora adusta, Phyllosticta ovalifolii): Brown or dark colored spots occur on the leaf. Generally not serious. Usually occurs during very rainy seasons or more frequently in overcrowded, poorly aerated plantings. Most prevalent just before normal fall of old leaves.

Sooty Mold: A blackish growth occurs on the leaves. Commonly follows white fly infestation. Control white flies.

Gall (fungus - Phomopsis sp.): Galls develop rapidly reaching a diameter of one and one-half inches within five and one-half months. These galls are commonly mistaken for bacterial galls. Prune out the galls.

NANDINA

Nandina domestica

Leaf Spots (fungi - Cercospora nandinae, Glomerella cingulata): Small, circular spots develop on the foliage. Lesion is brown-to-dark brown with reddish halo. Fungicides not required. Sanitation will help reduce foliage infection.

Ring Spot (virus): Light green or yellow, irregular rings occur in leaves. Plants tend to be smaller and low in vigor. Destroy diseased plants. Propagate from healthy plants.

Iron Chlorosis (nutrient deficiency): Foliage becomes light green and finally a greenish yellow. The veins tend to remain green. Foliar applications of iron sulfate or copperas are usually effective if several applications are made when leaves are forming. For mid- or late season corrections of a severe deficiency apply iron chelates such as Geigy 330 Fe (Central Texas) or Geigy 138 Fe (West and Far West Texas).

PHOTINIA

***Photinia* spp.**

Leaf Spots (fungi - Entomosporium maculatum, Cercospora heteromeles, Phyllosticta heteromeles): Spray at 7-10 day intervals during wet or humid weather with a recommended fungicide. Expanded, brown and often zoned spots on the leaves. The leaf spots may be round-to-irregular, purple, then gray-to-brown with purple margins in which black specks can later be found.

Powdery Mildew (fungi - Podosphaera leucotricha and Sphaerotheca pannosa): Dust or spray with approved powdery mildew fungicide when mildew first appears. Repeat at 10-14 day intervals as long as necessary. Preventive sprays in following years would produce better control.

Fire Blight (bacterium - Erwinia amylovora): Leaves, stems and fruit are blighted by this bacterium. New shoots suddenly appear as if scorched by fire. Brown or blackened leaves cling to twigs. Slightly sunken, girdling, discolored cankers develop on twigs, branches and trunk. Follow the same control procedures recommended for pears.

PITTOSPORUM

Pittosporum tobira

Cercospora Leaf Spot (fungus - Cercospora pittospori): Small, angular-shaped, yellowish-to-brownish spots. Spots may coalesce to form large irregular spots. Fungus fruits in fawn colored patches on lower surface. Most prevalent on older leaves. May cause premature leaf drop. Control is best obtained by protective spraying with recommended fungicides.

Leaf Spots (fungi - Alternaria tenuissima, Phyllosticta sp.): Small, circular dark-brown necrotic spots surrounded by chlorotic areas. Generally affects shrubs in poor physiological condition. Control is best obtained by maintaining shrubs in good growing condition. Applications of a protective fungicide are effective.

Cotton Root Rot (fungus - Phymatotrichum omnivorum): Plants suddenly wilt and die. When plants are pulled from the soil, the outer bark is decayed. Rhizomorphs of the fungus may be observed.

Root Knot Nematode (fungus - Meloidogyne sp.): Plants usually lack vigor, are often stunted and pale yellow. Plants may wilt quickly under moisture stress. Roots have small-to-large swellings, galls or knots. These are generally spherical and cannot be easily removed from the roots.

Virus: Three virus diseases, a mosaic on the leaves, one which causes a variegation of the leaves and one which causes rough bark have been reported on *Pittosporum*.

POINSETTIA

Euphorbia pulcherrima

Gray Mold (fungus - Botrytis cinerea): This disease is especially serious on double varieties. The inflorescence, leaves and stems are attacked. A gray mold occurs on the inflorescence causing a blasting and browning of the flower clusters and colored bracts. Water-soaked lesions will appear on leaves and stems. In the later stages of the leaf and stem disorder, the lesions become dry and crisp.

Cotton Root Rot (fungus - Phymatotrichum omnivorum): Cotton root rot is restricted to extreme South Texas where Poinsettias are planted in landscapes. In most areas of Texas, Poinsettias will not survive winter temperatures.

Scab (fungus - Sphaceloma poinsettia): A very common disease of poinsettia, particularly in South Texas. The fungus attacks leaves and stems, starting out as conspicuous, raised lesions or cankers on diseased stems. The cankers are at first white, later turning gray. Frequently, these cankers unite to cover irregular raised areas which give the plants an unsightly appearance. The disease cannot be controlled once it is well established on plants. Severely diseased plants should be cut back and the new growth should be protected with a fungicide.

Black Leg (fungus - Rhizoctonia solani): This disease can affect poinsettias at any stage of growth, but it is usually more prevalent right after the plants are potted. Stems affected by Rhizoctonia become dark-brown, water-soaked, with lesions that extend above and below the soil line. The lesion may encircle the stem, causing the stem to become weak. As a result of the black leg condition, the foliage becomes pale yellow and some of the leaves fall off. If roots and stems become severely infected, the plants die suddenly. Drenching the soil with a fungicide around the base of the plants may help in checking this disease.

Pythium Root Rot (fungus - Pythium sp.): Roots and basal portions of the stem turn brown and soft. Foliage may wilt rapidly. Pythium root rot is a major problem in commercial poinsettia production. Control measures consist of fungicide treatment of soil mix, fungicide drenches, acquisition of clean planting stock, and using a potting medium that drains well.

PYRACANTHA

***Pyracantha* spp.**

Scab (fungus - Fusicladium pyracanthae): Affects both the foliage and berries. It first shows up on foliage as small, greenish-yellow spots which later turn black. The berries are marked by small, black scabby spots. The disease is favored by long periods of intermittent rains and mild temperatures. Primary infection is from the previous year's foliage. Control is obtained by following a regular spray program and sanitation through removal of plant residues.

Fire Blight (bacterium - Erwinia amylovora): Causes twig dieback and blossom blight. Infected blossoms wilt and turn black rapidly. The disease progresses from the blossoms into the vascular system where it causes extensive twig death. The foliage does not abscise due to the rapid rate of death. Twig blight is common and can result in dieback ranging up to 12-24 inches of twig length. Dark sunken areas or cankers are formed on the twigs. Bacteria overwinter in the cankers. In the spring, bacteria ooze from the cankers and are carried by wind, rain and insects to healthy foliage. Shrubs which have fire blight should be pruned to remove infected material. Make cuts four inches below visible cankers. All pruning equipment should be sterilised between cuts with a one part household bleach to nine parts of water. Avoid high nitrogen levels.

Sooty Mold: Black, powdery mold in patches or covering entire upper leaf surfaces. A non-parasitic growth resulting from infestations of aphids or other sucking insects. Use effective insect control.

Root Rots (fungus): Tip burn and/or marginal burn of leaves, often followed by branch dieback or death of entire plant. Frequently associated with poorly drained or waterlogged soils.

ROSE

***Rosa* spp.**

Black Spot (fungus - Marsonnina rosea - Diplocarpon rosae): Black spots ranging in size from one-eighth to one-half inch or greater in diameter form on leaves. Spots can occur on either leaf surface as well as petioles, twigs and canes. Infected leaves soon turn yellow and drop from the plant. This results in plants becoming weakened to the point where few blossoms are produced and plant life is endangered. Consistent use of protectant fungicides is necessary for black spot prevention in most areas. The fungus persists in fallen leaves and canes. Destroy all fallen leaves and prune infected canes severely during late winter.

Powdery Mildew (fungus - Sphaerotheca pannosa var. rosae): A white powdery growth occurs on leaves, buds and twigs causing them to be distorted and dwarfed. Young, tender growth is most susceptible. The disease is more likely to occur during cool, dry conditions and can spread rapidly since a complete life cycle can occur in 72 hours. Thousands of spores are produced on a single plant with each having the ability to cause disease. Varieties differ in their susceptibility. Use an appropriate fungicide during times when disease pressure is high.

Dieback and Canker (fungi - Crytosporella umbrina and others): Plants weakened by other diseases or adverse growing conditions may be more susceptible to this condition. Organisms invade the stems and cause cankers (dead tissue surrounded by living tissue) to form. These often girdle the stem and tissue past the point of infection is killed. Prune infected limbs and twigs being sure to cut well below the diseased tissue. Disinfect shears in formaldehyde or chlorine solutions between each cut. Keep plants free of other diseases.

Rust (fungus - Phragmidium sp.): Rust is an occasional problem that can cause serious defoliation. Small, raised, orange pustules occur on the lower leaf surface. corresponding yellow spots occur on the upper surface. Affected leaves that shed from the plant should be collected and destroyed. Use a suggested protectant fungicide as soon as the first evidence of disease is observed.

Crown Gall (bacterium - Agrobacterium tumefaciens): Woody galls form on lower stems and root tissue. If the gall encircles the lower stem or main roots, the plant may be killed. Do not plant nursery stock showing these symptoms. Setting existing plant in raised beds may prolong their life. Sterilising the soil prior to planting is helpful only if the organism is not introduced on the transplant.

Verticillium Wilt (fungus - Verticillium sp.): Plants become unthrifty and die due to plugging of the water and nutrient conducting vessels. Occurs mostly in West Texas. Compost or sterilise gin trash if it is being used as mulch.

Root Knot Nematodes (nematode - Meloidogyne hapla): Knots or galls occur on smaller roots limiting the development of the root system. Do not plant stock that is infected. Sterilise soil if plants must be planted in infested areas.

Cotton Root Rot (fungus - Phymatotrichum omnivorum): Infected plants die suddenly. Roots may be covered with buff brown strands of fungal growth. Do not replant roses in the same area without soil sterilisation.

Virus Diseases: Roses are susceptible to several virus diseases, none of which can be controlled in an infected plant. The more common viruses are as follows:

— *Rose mosaic* :Variable yellow and green patterns occur in infected leaves. Some patterns may appear to be rather artistic. It appears to be spread from plant to plant only by graft transmission. No serious damage occurs to individual plants; however, grafting or budding should not be done from infected plants.

— *Spring Dwarf* : Symptoms are noticed when leaves first emerge in the spring. Leaves are balled, short and curved. Leaves have cleared veins.

WISTERIA

Wisteria spp.

Crown Gall (bacterium - Agrobacterium tumefaciens): This soil-borne bacterium will sometimes infect wisteria, causing galls or swellings on the main roots or stems. Inspect plants before buying and discard any plants with knots, galls, or swellings on the roots.

Leaf Spots (fungi - Phyllosticta wisteriae, Septoria wisteriae): Different fungi may cause leaf spots on wisteria. Pick off and destroy spotted leaves. Fungicide applications are normally not necessary.

Powdery Mildew (fungus -Erysiphe cichoracearum): A white-grayish mold develops on the surface of the leaves. Spray with a protective fungicide if the disease occurs frequently.

Cotton Root Rot (fungus - Phymatotrichum omnivorum): Wisteria is moderately susceptible to cotton root rot.

DISEASES OF ROSE PLANT

Black Spot symptoms

Black Spot (fungus - Marsonnina rosea - teleomorph Diplocarpon rosae): Black spots ranging in size from one-eighth to one-half inch or greater in diameter form on leaves. Spots can occur on either leaf surface as well as petioles, twigs and canes. Infected leaves soon turn yellow and drop from the plant. This results in plants becoming weakened to the point where few blossoms are produced and plant life is endangered. Consistent use of protectant fungicides is necessary for black spot prevention in most areas. The fungus persists in fallen leaves and canes. Destroy all fallen leaves and prune infected canes severely during late winter.

Brand and Common Canker symptoms

Brand and Common Cankers (fungi - Coniothyrium spp.): Plants weakened by other diseases or adverse growing conditions may be more susceptible. Organisms invade stems and cause cankers (dead tissue surrounded by living tissue) to form. These often girdle the stem and tissue past the point of infection is killed. Cankers are light brown in color. Prune infected limbs and twigs being sure to cut well below the diseased tissue. Keep plants free of other diseases.

Crown Canker symptoms

Crown Canker (fungus - Cylindrocladium scoparium): Black, watersoaked cankers develop at the base of plants. They usually go unnoticed until canes begin to dieback. Crown canker is primarily a problem on greenhouse roses or roses, like miniature types, propagated in containers. Sanitation is the key to control. Chemicals have not been highly effective.

Crown Gall symptoms

Crown Gall (bacterium - Agrobacterium tumefaciens): Woody

galls form on lower stems and root tissue. If the gall encircles the lower stem or main roots, the plant may be killed. Do not plant nursery stock showing symptoms. Resetting galled plants in raised beds may prolong their life. Sterilising the soil prior to planting is helpful only if the organism is not introduced on the transplant.

Downy Mildew symptoms

Downy Mildew (fungus - Peronospora sparsa): A highly destructive disease on rose, causing severe and rapid defoliation. The fungus is active only under cool, damp conditions. Leaf symptoms vary from angular blotches (yellow, purple to brown) to a scorch like burn. Downy mildew has not been a problem in landscapes in Texas. It has on occasion developed in commercial operations where roses are grown temporarily in containers. Regular overhead watering and thick growth from close spacing create ideal conditions for this disease. Preventive fungicides should be applied when weather is ideal for disease development.

Mosaic Virus symptoms

Mosaic Virus (virus - Prunus necrotic ringspot, Apple mosaic and others): Variable yellow and green patterns occur in leaves. Some patterns may appear to be rather artistic. Symptoms on foliage are most evident in spring. As the season progresses symptoms often fade. These viruses are not lethal but can reduce vigor and overall performance. There is no control after plants are infected. The primary means of transmission for these viruses is through grafting. No insect vector has been found.

Powdery Mildew symptoms

Powdery Mildew (fungus - Sphaerotheca pannosa var. rosae): A white powdery growth develops on leaves, buds and twigs causing them to be distorted and dwarfed. Young, tender growth is most susceptible. The disease is more likely to occur during cool, dry conditions and can spread rapidly since a complete life cycle can occur in 72 hours. Thousands of spores are produced on a single plant with each having the ability to cause disease. Varieties differ in their susceptibility. Use an appropriate fungicide during times when disease pressure is high.

Root Knot Nematodes symptoms

Root Knot Nematodes (nematode - Meloidogyne hapla): Knots or galls occur on smaller roots limiting the development of the root system. Do not plant stock that is infected. Raised beds, with heavy pine bark mulch, seem to deter root knot nematode activity.

Rose Rosette symptoms

Rose Rosette (cause not confirmed): This disease has all the characteristics of a virus infection. The disease agent is transmitted by a microscopic sized eriophyid mite. Symptoms include distorted, small leaves, short internodes, red pigmentation of foliage and excessive thorniness. Infected plants generally die within one year. Hybrid tea roses are poor hosts for the mite. Rose rosette has not been identified in landscape plantings in Texas but has been diagnosed in multiflora rose.

Rust Symptoms

Rust (fungus - Phragmidium spp.): Rust is a rare problem but can cause serious defoliation. Small, raised, orange pustules occur on the lower leaf surface. Corresponding yellow spots occur on the upper surface. Affected leaves that shed from the plant should be collected and destroyed. Use a protectant fungicide as soon as the first evidence of disease is observed.

REFERENCES

Forsberg, Junius L., *Diseases of Ornamental Plants*, University of Illinois. Urbana-Champaign, Illinois, 1975.

Grimm, William C., *How To Recognize Shrubs*, Castle Books, NY, 1966.

Johnson, Warren T. and Lyon, Howard H., *Insects That Feed on Trees and Shrubs*, Cornell University Press. Ithàca, New York, 1976.

Melle, P. J. Van, *Shrubs and Trees for the Small Place*, Doubleday and Company, Inc. Garden City, New York, 1955.

Sinclair, Wayne A.; Lyon, Howard H.; Johnson, Warren T., *Diseases of Trees and Shrubs*, Cornell University Press. Ithaca, New York, 1996.

10

Diseases of Cereal Crops

DISEASES OF BARLEY

Barley Yellow Dwarf Symptoms

Barley Yellow Dwarf Virus (virus): Yellowing of leaves is the first visible sign that a plant is diseased. When infection occurs in a plant during the seedling stage, the leaves turn a bright golden yellow, usually beginning at the tips and progressing back along the edges. In very susceptible varieties, plants infected in the seedling stage are severely dwarfed and may grow less than six inches tall. They produce more tillers than healthy plants but form no heads. Several species of aphids can carry the virus from infected grasses to barley. Use resistant varieties.

Basal Glume Blotch Symptoms

Basal Glume Blotch (bacterium - Pseudomonas atrofaciens): Basal glume blotch on barley is caused by a bacterium. It is caused by the same bacterium that occurs commonly on wheat. The symptom of basal glume blotch is dull brownish-black, discolored area found at the base of each glume covering the kernel. The discoloration is more pronounced on the inside than on the outside of the diseased glume. The base of a diseased kernel shows a discoloration varying from faint brown to charcoal black with the color depending upon the severity of the attack.

Leaves affected by this disease organism show small, dark, water-soaked spots. These spots tend to enlarge and turn yellow and finally brown as the tissue dies. The use of clean and treated seed will reduce the severity of this disease.

Leaf Blotch Symptoms

Leaf Blotch (fungus - Septoria passerinii): The spots characteristic of leaf blotch are elongated and yellowish-brown, and their ends have indefinite margins. Numerous dark-brown fruiting bodies develop in rows between the veins in the dead, straw-colored parts of the spots. Control is the same as for spot blotch.

Leaf Rust Symptoms

Leaf Rust (fungus - Puccinia hordei): Small, round yellow or yellowish-brown pustules appear on the leaves. Infected leaves eventually die. Leaf rust of barley does not infect oats or wheat. The development of resistant varieties is currently the only practical means of control.

Nematode Symptoms

Nematodes: Lesion, spiral, stubby root and stunt nematodes feed on the roots and can result in poor root development. Plants will be stunted and in most cases show an off-color as a nutritional deficiency. Frequent turning of the soil during summer months tends to lower populations.

Net Blotch Symptoms

Net Blotch (fungus - Pyrenophora teres): Net blotch of barley is principally a leaf disease. The spots are internal with a characteristic netting. The netted pattern, formed by the arrangement of brown pigment in transverse and longitudinal lines, can best be seen if the leaf is held against the light. The elongated, brown-netted areas finally cover most of the leaf and destroy it usefulness. The fungus causing net blotch may be carried on the seed or infection may come from old straw or stubble. Crop rotation, plowing old straw and stubble, and seed treatment will aid in prevention of this disease. Avoid susceptible varieties.

Powdery Mildew Symptoms

Powdery mildew (*Erysiphe graminis*); Gray, fluffy threads of the fungus can be observed on the upper surface of the leaves and leaf sheaths. The fungal strands penetrate and invade the leaf tissue. As the disease progresses, the leaf turns yellow and is gradually killed. Spores of the fungus are carried from plant-to-plant by wind currents. While the disease can be controlled with fungicides, such controls may not be economical.

Smut Symptoms

Smut (fungi - Ustilago hordei, U. nigra, U. nuda): Three smuts occur on barley in Texas, but with seed treatment fungicides now available, each can be controlled. Covered smut (*U. hordei*) is noticed at heading time. Hard, black masses of smut, each covered with a grayish membrane, are found in place of kernels. Semi-loose smut (*U. nigra*) destroys the floral parts and entire head. Spores of the nigra smut are carried within the husk covering or on the seed. Loose smut (*U. nuda*) is similar to semi-loose smut except the smut fungus is internally seed-borne.

Spot Blotch Symptoms

Spot Blotch (fungus - Bipolaris sorokiniana): Spot blotch infection starts as dark brown-to-black spots on sheaths that cover the young shoots. The infection progresses inward and sometimes kills the seedlings below the surface of the ground, but more often it kills the seedlings after emergence. Diseased seedlings are dwarfed and tiller excessively. The leaf spot has round to oblong dark brown lesions. Spots may coalesce to form blotches. As crown rot, the disease develops at or near the soil line. Crown roots may be rotted and tillers killed. The pathogen can be soil- or seed-borne, and may be carried on crop residue. Seed treatment with a fungicide reduces losses from seedling infection. Clean plowing of crop residue and crop rotation will aid in control.

CORN

Bacterial Leaf Blight Symptoms

Bacterial Leaf Blight (bacterium - Erwinia Stewartii): This disease is also called Stewart's bacterial wilt, Stewart's leaf blight and maize bacteriosis, and is rarely found in Texas. Leaves show linear, pale green-yellow streaks with irregular margins that parallel veins. These streaks become dry and brown. Plants produce premature, bleached, dead tassels. Severely infected plants form cavities in the pith of the lower stalk and the bacteria spread through the vascular system into the kernels. Chewing insects, such as flea beetles, wireworms, root worms and May beetles can spread the bacterium. Field corn is more resistant than sweet corn. Control by using resistant hybrids and control chewing insects.

Brown Spot Symptoms

Brown Spot (fungus - Physoderma maydis): Brown spot is generally not a serious problem. Symptoms are small, round, yellow spots on leaves and sheathes that appear in bands across the leaf. Lesions turn brown and blend together to form larger blotches . Infected leaf tissue decomposes, releasing brown spores. Stalks often break and lodge.

Common rust

Common Rust (fungus - Puccinia sorghi): The fungus erupts from the leaf tissue in round to oval areas called pustules, with cinnamon brown spores . These spores eventually turn black. Rust is favored by cool temperatures and high humidity. Older tissue is generally resistant. There are numerous physiologic races of this fungus. Control disease with slow rusting resistant hybrids. Fungicides are usually not economically practical. High summer temperatures usually reduce the disease.

Common smut

Common Smut (fungus - Ustilago maydis): All parts of the plant are susceptible. Galls are covered with a white membrane. The inside of the gall turns into a mass of powdery black spores. Early infection may kill small plants but this is rare. A gall on the lower part of the stalk can make the ears small or non-existent. Dry weather with 78° - 95° F is favorable. High nitrogen or heavy manure is also favorable. Injuries from blowing sand, hail damage, cultivation, or buggy-whipping will increase smut. Avoid mechanical injury, avoid highly susceptible varieties, and maintain balanced fertility

Crazy top

Crazy Top (fungus - Sclerophthora macrospora): Symptoms vary with time of infection and degree of colonisation. First symptoms are excessive tillering, rolling, and twisting of upper leaves. The tassel proliferates until it is just a mass of leaves, hence the name crazy top . Leaves may also be stunted with yellow stripes. Flowering before plants reach the four- to five-leaf stage is usually a factor. Losses are usually confined to small areas. Avoid wet fields with poor drainage.

Ear and kernel rots

Ear and Kernel Rots (fungi - several species): Corn ear and kernel rots are a greater problem in areas with high rain fall from silking to harvest. Symptoms range from pink mold on ears with Fusarium to green mold with Tricoderma. Moldy corn may contain toxic materials. Bird and insect damage to ear and stalks along with lodging will increase stalk rots. Rots are reduced in ears that mature hanging downward. Yield, quality, and feed value can be reduced. Controls include resistant hybrids, early harvest, and proper storage below 15% moisture.

Head smut

Head Smut (fungus - Sporisorium reilianum): This soil-borne fungus attacks both corn and sorghum. It appears on the ear and tassels. Tassel infection may be from individual spihelets or a large mass of black spores. Usually if the tassels are smutted, all the ears will be smutted. Controls include resistant varieties or hybrids and rotation

Nematodes

Nematodes: In general, nematodes decrease the efficiency of the root system, resulting in reduced growth, yellowing and reduced yields. More than 40 species have been reported on corn. The main symptoms on corn are lack of vigor and dark, discolored areas on the roots. Rotate with non-host plants and use chemical infurrow treatments where practical.

Northern corn leaf blight

Northern Corn Leaf Blight (fungus - Exserolilium turcicum): Long, elliptical, tan lesions develop first on older lower leaves. It progresses up the plant and resembles frost or drought injury. Lesions form on husks but ears are not infected. Lesions on resistant hybrids have a yellow margin. It overwinters on leaves, husks, and other plant parts in debris. In wet weather, black spores are produced on the lesions that are wind-blown long distances. Secondary spread is common within and between fields. Moderate temperatures, heavy dews, and frequent showers are favorable. High losses will occur with infection before silking. Infection six weeks after silking will cause minimal loss. Control with resistant hybrids. Check for approved fungicides. Section 18 emergency exemptions for Tilt have been available in the past.

Seed rots/seedling blight

Seed Rots and Seedling Blight (fungi - several species): Soil and seed-borne fungi attack corn seed causing seed rots and seedling blight. These fungi are favored by cold (50 - 55° C) wet, poorly drained soils. Factors that affect disease severity include genetic resistance, seed quality, planting depth, and soil type. Symptoms include yellowing and wilt of leaves, seed rot, damping-off of seedlings, seedling wilt, and root rot. Controls include planting high quality damage-free seed, planting in warm, moist soil, correct placement of infurrow pesticides and fertiliser, and seed treatment fungicides.

Sorghum downy mildew

Sorghum Downy Mildew (fungus - Peronoscherospora sorghi): Symptoms are yellow, stunted plants with occasional white striped leaves. Infection is systemic. Infected plants have narrow, more erect leaves. Plants will have bushy tassels and downy white growth will appear on both leaf surfaces. Tolerant plants show some symptoms but still produce seed. Some varieties show long, narrow, yellow lesions. Spores can blow from plant to plant or field to field. Other spores survive for years in the soil. Use resistant hybrids and systemic fungicide seed treatment.

Southern Corn Leaf Blight

Southern Corn Leaf Blight (fungus - Bipolaris maydis): This was a minor disease of corn for many years, with no economic effect on yield. In 1970, a highly virulent strain called Race T appeared on corn hybrids with Texas male sterile cytoplasm. Losses were severe. Symptoms of Race lesions are spindle-shaped with yellow-green halos. They will later have dark, reddish-brown borders and occur on leaves, stalks, leaf sheaths, husks, and shanks. Cob Rot ear occurs as well as ear drop. The fungus overwinters in corn debris and on seed. Wind and splashing water spread the spores rapidly in the field under ideal conditions, cycling in about 72 hours. Control with resistant hybrids

Southern corn rust

Southern Corn Rust (fungus - Puccinia polysora): Southern rust develops more rapidly and is more destructive to leaf tissue than common rust, but it occurs primarily along the Coastal Bend as far north as Hempstead. Southern rust pustules are smaller, more round and orange in color than common rust. The many pustules that form cause the leaf to turn yellow and die. Yields are reduced by heavy infection at ear filling. Unlike common rust, southern rust is favored by high temperature. Corn hybrids vary in susceptibility to this disease. Resistant hybrids should be planted. Chemical control is available to growers through mancozeb fungicides.

Stalk rots

Stalk Rots (fungi - several species): Stalk rots are the world's most destructive corn diseases. They are caused by a complex of fungi and bacteria that attack the stalks near maturity. Stalk rots are favored by conditions that encourage heavy kernel set followed by late season stress, such as leaf blights, extended cloudiness, heavy plant stands, drought, hail, and low K with high N. Most stalk rots are favored by dry conditions early and warm, wet weather after silking. Symptoms include destruction of pith in the stalk and lodging. Interior discoloration ranges from pink with Gibberella to gray with charcoal rot. Yield reductions of 10 - 20% are common with susceptible hybrids. Losses occur from both poorly filled ears and harvest losses due to lodging. Controls include using disease-free seed of resistant hybrids, crop rotation, balanced N and K fertility, sanitation, stress reduction such as pest control, and reduced plant populations.

Maize dwarf mosaic virus

Maize Dwarf Mosaic Virus (virus - maize dwarf mosaic virus (MDMV)): This is the most common and most damaging of corn viruses in Texas. Youngest leaves have a light to dark green mottle or mosaic in narrow streaks along veins. Plants are stunted, excessively tillered, with multiple ear shoots and poor seed set. Early infection can lead to root and stalk rots and death. The virus is transmitted mechanically by at least 12 different aphids. Symptoms appear 30 days after emergence. Many wild and cultivated grasses serve as hosts for MDMV. Johnsongrass is considered the main overwintering reservoir. Use resistant hybrids and control Johnsongrass.

DISEASES OF OATS

Crown or Leaf Rust

Crown or Leaf Rust (fungus - Puccinia coronata f. sp. avenae): This fungus overwinters on volunteer and wild oats. Alternate hosts of this rust are species of Rhamnus.The disease appears as tiny, round, yellowish-red pustules on either surface of the leaf blade. These pustules produce tiny yellow-red spores that are spread to other plants by wind currents and germinate in free moisture from dews and rains. Crown rust, like other rusts, is caused by a fungus that attacks the leaves and leaf sheath of oats and related grasses. Grazing the infected crop is a suitable means of salvaging value from the crop when damage is heavy. Some varieties have resistance.

Halo blight

Halo Blight (bacterium - Pseudomonas coronafaciens): Lesions or spots appear on leaf blades as oval to oblong water-soaked spots that are buff to light brown in color. A light-yellow zone forms a halo around the brown lesion. The bacterium enters the leaf tissue through stomata or an insect puncture. Suggested controls are seed treatment, crop residue management and rotation.

Leaf blotch, victoria blight, culm rot

Leaf Blotch, Victoria Blight, Culm Rot (fungi - Drechslera avenacea, Bipolaris victoriae, Bipolaris maydis): Three species of fungi cause economically significant diseases of oats. With leaf blotch, oblong, linear blotches appear on the leaves. The blotches are light-yellow at first, later turning red to brown. Victoria blight is a seedling blight and sometimes seedlings are killed soon after they emerge. In other cases, plants are stunted, leaves turn red and lower leaves eventually die. Culm rot is a basal stem and root rot. Seed treatment and crop rotation will help prevent this disease.

Powdery mildew

Powdery Mildew (fungus - Erysiphe graminis): This fungus disease appears as patches of white, fluffy growth on the lower leaves and leaf sheaths. No practical control measure is available. Grazing excessive growth will allow light and air to keep the plants dry, thus reducing the severity of powdery mildew.

Smuts

Smuts (fungi - Ustilago avenae and Ustilago kolleri): Loose smut (*Ustilago avenae*) and covered smut (*Ustilago kolleri*) are two of the most common and widespread of the smut diseases of oats. The smut balls of loose smut form in the oat panicle usually destroying most or all of the seed and hulls. Wind and rain usually dislodge most of the spores soon after they appear and infected panicles are often inconspicuous in the field. The smut balls of covered smut do not ordinarily destroy the outer hulls and the spore masses remain more or less intact inside the hulls until threshing time. Seed treatment with recommended fungicide will prevent both type smuts

Stem rusts

Stem Rust (fungus - Puccinia graminis f. sp. avenae): The pustules of stem rust are usually oblong and dark reddish-brown. The epidermis is ruptured and often pushed up around the edges and through the center of the pustules so that it is easily visible. The dark color of the stem rust pustule distinguishes it from the light, orange-yellow crown rust.

DISEASES OF SORGHUM

Anthracnose

Anthracnose (fungus - Colletotrichum graminicola): The anthracnose fungus damages foliage and stems of grain sorghum. On susceptible hybrids, the stem holding the head (peduncle) becomes infected and a brown sunken area with distinct margins develops. When infected stems are cut lenghtwise with a knife, one can see that the fungus has penetrated the soft pith tissue and caused brick-red discolorations. This peduncle infection inhibits the flow of water and nutrients to the grain causing poor grain development. The fungus also invades individual grains and the small branches of the panicle. Rapid and severe yield loss can result from panicle and peduncle infections. Leaf lesions are small, elliptical to circular, usually less than 3/8-inch in diameter. These spots develop small, circular, straw-colored centers with wide margins that may vary in color from reddish to tan to blackish purple. The spots may coalesce to form larger areas of infected tissue.

Bacterial Diseases

Charcoal rot

Charcoal Rot (fungus - Macrophomina phaseolina): Grain sorghum plants affected by the charcoal rot fungus fail to fill grain properly and may lodge in the latter part of the season. Infected stalks show an internal shredding at and above the ground line. This can be observed by splitting the stalk and noting the deteriorated soft pith tissue leaving the tougher vascular strands. Fungal structures (sclerotia) can be observed in the affected tissue which appears as though it has been dusted with black pepper. Another type of stalk rot (*Pythium* sp. and *Fusarium* sp.) may show the shredded condition but the black specks (sclerotia) will be lacking. Conditions under which charcoal rot is favored include stressful hot soil tempertures and low soil moisture during the postflowering period. Host plants are usually in the early-milk to late-dough stage when infection occurs. The fungus is common and widely distributed in nature. Avoiding moisture stress, proper management of crop residue, crop rotation, avoiding excessive plant populations, balancing nitrogen and potassium fertility levels, and growing drought-tolerant, lodging-resistant hybrids represent the best means of control.

Covered kernel smut

Covered Kernel Smut (fungus - Sporisorium sorghi): This smut disease was once quite destructive but is seldom seen now because most seed is chemically treated. The disease destroys all kernels in a head and replaces them with a cone-shaped gall or may affect only portions of a panicle. At harvest time, these galls are broken and spores contaminate the outer surface of other kernels. This disease is controlled by use of chemical seed treatment, use of clean seed and planting resistant hybrids.

Crazy top downy mildew

Crazy Top Downy Mildew (fungus - Sclerophthora macrospora): This fungal disease can be troublesome in low lying areas that become flooded. Infected plants have thick, stiff, twisted, pale green leaves with bumpy surfaces. The leaves often turn downward, and the plants produce many shoots or suckers giving a bunchy appearance. Infected plants do not produce heads or produce a proliferation of leafy tissue in place of the head. Wild and cultivated grasses can serve as sources of inoculum. There are differences in susceptibility among grain sorghum hybrids but these differences are not significant. Installing drainage structures and diverting water movement to avoid flooding is the most reasonable suggestion for control at this time.

Foliage diseases/fungal

Foliage Diseases Caused by Fungi: A number of fungal organisms cause foliage infection and may become severe under certain conditions. Occurrence has seldom been consistent or damaging enough to warrant the development of specific control practices. No fungicides have been cleared for use on sorghum foliage. Hybrids vary in their susceptibility to these diseases and resistance is available if their occurrence becomes a problem

Fusarium stalk rot

Fusarium Stalk Rot (fungus - Fusarium moniliforme): Like charcoal rot, *Fusarium* stalk rot usually develops on mature to nearly mature plants that have been subjected to some form of stress. Infection takes place at the base of the plant and produces discoloration in the stalk. When shredding of the vascular area occurs from this organism, black fungal bodies (sclerotia) are not present as they are with charcoal rot. Avoiding stress problems by proper use of cultural practices is the best approach to control.

Head smut

Head Smut (fungus - Sporisorium reilianum): This disease is characterised by the large, dark-brown smut galls that emerge in place of the panicle . The gall is first covered with a whitish membrane which soon breaks and allows spores to be scattered by the wind. Plants become infected while in the seedling stage but evidence of infection is not apparent until heading time. The smut gall produces thousands of spores which become soil-borne and initiate systemic infection of seedlings in subsequent years. Different races of the fungus exist which may result in a sorghum hybrid being resistant in one area but not another. New sources of resistance have been found and growers should utilise resistant hybrids to avoid losses from this disease.

Loose kernel smut

Loose Kernel Smut (fungus - Sporisorium cruenta): Galls formed by loose kernel smut are long and pointed and the thin membrane covering them usually breaks soon after galls reach full size. This disease presents no immediate problem because the control measures mentioned for covered kernel smut have virtually eliminated occurrence of this disease.

Maize dwarf mosaic virus

Maize Dwarf Mosaic Virus (MDMV): maize dwarf mosaic is a virus disease that occurs over all the sorghum producing areas of Texas. Its ability to cause damage is dependent on the presence of an overwintering virus host (mainly Johnsongrass), aphid populations to facilitate virus transmission and the susceptibility of the hybrid being grown. Affected plants have mottled (light green blotchiness) terminal leaves. These alternate light- and darker-green areas in the leaf can be more easily seen when held between the viewer and a light source. Observers should always look at the newest leaves for the most severe symptoms. Highly susceptible hybrids are stunted with chlorotic symptoms in the upper leaves and suffer significant yield losses. Some hybrids produce a red leaf symptom when plants are infected and when night temperatures are below 55 degrees F. Use tolerant hybrids and control Johnsongrass in and around the field to manage this disease.

Root rot complex

Root Rot Complex (fungi - Fusarium moniliforme, Pythium sp. and others): Several fungi are involved in producing a root rot condition of grain sorghum. One or more of the causal fungi may be involved depending on conditions and organisms present in the soil. Each organism produces distinct symptoms, but identification becomes more complex when other factors are involved. Rotation with non-related crops will lower the population of pathogenic organisms present in the soil.

Rust

Rust (fungus - Puccinia purpurea): Rust appears on leaves as small raised pustules or blisters that rupture and release many reddish-brown spores. These pustules occur on both the upper and lower leaf surfaces. This disease usually appears when plants near maturity and infection is confined primarily to mature leaves. Grain yield losses are usually not serious and occurrence of the disease is sporadic. Forage sorghum yields may be affected most. The rust fungus also attacks Johnsongrass and overwinters on this host in southern production areas.

SEEDLING DISEASES

Seed Rots and Seedling Diseases (fungi - Rhizoctonia solani, Fusarium sp., Pythium sp. and others): Seed rots and seedling diseases are more prevalent when seed is planted in cool, wet soils, and especially when packing rains seal the soil surface. Seedling emergence and growth is slowed under such conditions, and several soil-borne fungi are able to infect the tender seedlings. Other factors such as poor seed quality or improperly placed fertiliser or herbicide often contribute to the problem. Continuous growth of one crop over a period of years may favor buildup of organisms that cause seedling disease. Use of high quality seed treated with a recommended seed protectant is important. Precision planting to insure proper depth in a well-prepared seedbed is very helpful.

Sorghum downy mildew

Sorghum Downy Mildew (fungus - Peronosclerospora sorghi): Young, systemically infected plants have light green to yellowish stripes lengthwise in the leaves often with a grayish-white downy fungal growth consisting of numerous tiny spores on the lower surface of the leaf opposite the pale striped areas. Soil-borne spores cause systemic infection of the young seedlings. These systemically infected plants will not produce heads. Later infections may parially or completely inhibit grain formation. Plants can also be infected by air-borne spores from the underside of infected leaves. These infections result in localised leaf lesions. Grain sorghum hybrids vary considerably in their susceptibility. Growers should select adapted, high producing hybrids that have resistance to this fungus. Growing highly susceptible grain or forage sorghum builds up populations of the organism in soil, making the problem more severe even in resistant hybrids. Use seed treated with a systemic fungicide containing metalaxyl and resistant hybrids to control this disease.

DISEASES OF RICE

Bacterial Leaf Blight

Bacterial Leaf Blight (bacterium - Xanthomonas campestris pv oryzae: This disease was identified for the first time in Texas in 1987. The disease is very destructive in Asia, but at this time it appears that the strain of the bacterium in Texas is not very aggressive and causes negligible losses. The first symptom of the disease is a water soaked lesion on the edges of the leaf blades near the leaf tip. The lesions expand and turn yellowish and eventually grayish-white. High rainfall with strong winds are thought thought to provide conditions for the bacteria to multiply and enter the leaf through injured tissue. The bacterium has been isolated from clubhead grass (*Leersia hexandra*) collected form rice fields in Texas. Fall plowing or rolling of stubble to hasten decay of the rice debris should help to manage the disease by destroying the tissue in which the bacterium is maintained.

Brown Leaf Spot

Brown Leaf Spot (fungus - Bipolaris oryzae): This disease, previously called *Helminthosporium* leaf spot, is common in Texas. Most conspicuous symptoms of the disease occur on leaves and glumes of maturing plants. Symptoms also appear on young seedlings and the panicle branches in older plants. Brown leaf spot is a seed-borne disease. Leaf spots may be evident shortly after seedling emergence and continue to develop until maturity. Leaf spots vary in size, are typically 1/8 inch in diameter, and are circular to oval in shape. The smaller spots are dark brown to reddish brown, and the larger spots have a darkbrown margin and reddish brown to gray centers. Damage from brown spot is particularly noticeable when the crop is produced in nutritionally deficient or otherwise unfavorable soil conditions. Significant development of brown spot is often indicative of a soil fertility problem. Brown spot may be reduced by balanced fertilisation, crop rotation, and the use of high quality planting seed. Foliar fungicides are not economical for controlling brown leaf spot on most commercial long grain varieties. Rice seed with infected glumes can result in diseased seedlings. Seed treatment fungicides reduce the incidence and severity of seedling blight caused by this fungus.

Blast

Blast (fungus - Pyricularia grisea): This disease can cause serious losses to susceptible varieties during periods of blast favorable weather. Depending on the part of the plant affected, the disease is often called leaf blast, rotten neck, or panicle blast. The fungus produces spots or lesions on leaves, nodes, panicles, and collar of the flag leaves. Leaf lesions range from somewhat diamondshaped to elongated with tapered, pointed ends. The center of the spot is usually gray and the margin brown or reddish-brown. Both the shape and color of the spots may vary and resemble those of the brown leaf spot disease. Blast differs from brown leaf spot in that it causes longer lesions and develops more rapidly.

The blast fungus frequently attacks the node at the base of the panicle and the branches of the panicle. If the panicle is attacked early in its development, the grain on the lower portion of the panicle may be blank giving the head a bleached whitish color, giving the term "blasted" head or rice "blast". If the node at the base of the panicle is infected, the panicle breaks causing the "rotten neck" condition. In addition, the fungus may also attack the nodes or joints of the stem. When a node is infected, the sheath tissue rots and the part of the stem above the point of infection often is killed. In some cases, the node is weakened to the extent that the stem will break causing extensive lodging.

Blast generally occurs scattered throughout a field rather than in a localised area of the field. Late planting, frequent showers, overcast skies, and warm weather favor development of blast. Spores of the fungus are produced in great abundance on blast lesions and can become airborne, disseminating the fungus a considerable distance. High nitrogen fertilisation should be avoided in areas that have a history of blast. Control measures include early planting, avoiding excessive or high levels of nitrogen, proper flood management, resistant varieties, and fungicides. Varietal resistance is the most effective method of controlling rice blast. Some foliar fungicides can reduce the incidence of blast, but severe losses can occur on susceptible varieties even when fungicides are applied.

Kernal Smut

Kernel Smut (fungus - Neovossia barclayana): This disease causes losses in both yield and quality. The endosperm of the grain is attacked by the fungus causing either part or all of the starchy material to be replaced by a black mass of smut spores. The fungus does not destroy the seed embryo, and the diseased seed will germinate even if all the endosperm has been replaced. Release of the smut spores from within the kernel will cause a discoloration of hulls. Moisture causes the dark mass of spores to swell and break out of the hull. Smut is easily detected after a rain or in the early morning following a heavy dew.

Kernel smut is usually most severe on late planted rice. Usually, only a small percent of the kernels are affected, but monetary losses can be significant because of penalties levied on the grain at the dryer. Milled rice has a dull or grayish appearance when smutted grains are present in the sample. High rates of nitrogen increase the incidence of smut. Varieties differ in the incidence of smut observed in commercial plantings. Less smut is detected in most commercial semi-dwarf varieties.

Control measures include the use of semi-dwarf varieties in fields with a history of smut and to reduce reduce nitrogen rates and floodwater depths in fields where very susceptible varieties must be grown. The fungicide propiconazole is effective in suppressing kernel smut.

Leaf Smut

Leaf Smut (fungus - Entyloma oryzae): This is a minor fungal disease in which small slightly raised black spots develop primarily on the leaves. Raised spots or pustules break open releasing air-borne spores. Infection is often heavy enough to kill tips of leaves. Leaf smut occurs late in the growing season and causes little or no economic loss. No control measures are recommended.

Narrow Brown Leaf Spot

Narrow Brown Leaf Spot (fungus - Cercospora janseana): The disease varies in severity from year to year and usually becomes most severe as rice approaches maturity, causing premature ripening and yield reduction. Leaf spots are long, narrow, and cinnamon-brown. Premature leaf death will occur in severe cases. Late in the growing season, the fungus often attacks the sheath of the flag leaf causing the "brown blotch" or "net blotch" phase of the disease in which a large cinnamon brown lesion is formed and typically encircles the uppermost internode about an inch below the the base of the panicle.

Early maturing varieties tend to escape the major impact of the disease. There are differences in suseptibility among some rice varieties, however, due to buildup of certain races of the fungus, resistance does not remain reliable. Some foliar fungicides effectively suppress this disease and may be economical if other diseases are also controlled along with the narrow brown leaf spot.

Seedling Blight/Seed Decay

Seedling Blight and Seed Decay (fungus - Bipolaris oryzae, Pythium sp., Rhizoctonia solani, Achlya sp. and Sclerotium rolfsii): Seedling diseases cause spotty, irregular stands through seed decay, pre-emergence and post emergence disease. Seedling disease complex results from activity of various kinds of fungi, most of which grow on the kernels or hulls of seed rice or on soil organic matter. Fungi enter germinating rice seed or young seedlings and either injure or kill them. If infected seedlings emerge from the soil, they often die. Those that manage to survive are weakened and chlorotic in appearance. Damage is most severe on early seeded rice and deeply planted rice. Control measures include the use of high quality seed, an approved seed treatment, shallow seeding of early planted rice, and planting into warm soil.

Sheath Blight

Sheath Blight (fungus - Rhizoctonia solani): Sheath blight is perhaps the most important disease of rice in Texas. Initial symptoms usually develop as lesions on sheaths of lower leaves near the water line when plants are in the late tillering or early internode elongation stage of growth (approximately 10 - 15 days after flooding). These lesions usually develop just below the leaf collar as oval-to-elliptical, green-gray, water-soaked spots about 1/4 inch wide and 1/2 to 1 1/4 inch long

With age, the lesions expand and the center of the lesions may become bleached with an irregular tan-to-brown border. When humidity exceeds 95 percent and temperatures are in the range of 85-90 degrees F, infection spreads rapidly by means of runner hyphae to upper plant parts, including leaf blades, causing extensive, tan, irregularly shaped lesions with brown borders. Disease development progresses very rapidly in the early heading and grain filling growth stages during periods of frequent rainfall and overcast skies. Plants heavily infected at these stages produce poorly filled grain, particularly in the lower portion of the panicle. Additional losses result from increased lodging or reduced ratoon production due to infection of the culm and reduced carbohydrate reserves.

As plants senesce from maturity, lesions will dry and become grayishwhite to tan with brownish borders. Sclerotia, initially white but turning dark brown at maturity, are produced superficially on or near the lesions. Sclerotia are loosely attached and easily dislodge from the plant. Sclerotia are the primary means for fungus survival between crops. They survive long periods in the soil and will float to the surface of flooded rice fields in the subsequent rice crop, infect rice plants at the waterline and continue the disease cycle. Sclerotia can survive from one to several years in the soil. They can also attack several weed hosts and cause infection.

New varieties and changing cultural practices often combine many of the factors that favor disease development. In recent years, the wide acceptance of susceptible varieties, because of their high yielding potential, has contributed greatly to the rapid increase in sheath blight. In addition, the new varieties respond to heavy nitrogen applications in order to achieve their high yielding potential. Heavy applications of nitrogen predispose susceptible plants to attack by the sheath blight organism. Rotation with susceptible crops, such as soybean can also in crease the severity of sheath blight in succeeding rice crops.

Disease incidence may be reduced by planting less susceptible varieties. None of the long grain varieties currently grown in Texas show adequate field resistance. Excessive seeding rates and high nitrogen applications should be avoided in fields with a history of the disease. Grass and weeds should be controlled. Long-term rotations may reduce the incidence of sheath blight, but soybeans, sorghum, and many weeds are susceptible to *Rhizoctonia solani.* Excessive seeding rates and thick plant populations favor sheath blight development. In some cases, foliar fungicides may be economical for reducing sheath blight losses

Sheath Spot

Sheath Spot (fungus - Rhizoctonia oryzae): This disease is similar to the early stages of sheath blight and is caused by a related fungus. Small sclerotia initiate the disease on the outermost leaf sheath. A reddish-brown lesion develops on the plant near the waterline. During internode elongation, the sheath spot may be pushed up above the waterline because of this new growth. With age, the lesion enlarges slightly to assume an eliptical to irregular shape (1/2 to 1 inch long) with a distinct purple-brown border and a tan-to-straw colored center. The sheath spot fungus does not develop on the leaf blades as is the case with sheath blight. Rather it remains on the outer sheath only. Sometimes the fungus will cause a yellowing of the leaf attached to the sheath it has infected. Some reduction in yield may occur in tillers that are infected; however, it is generally minor and too few tillers are affected to result in a significant yield loss.

Stem Rot

Stem Rot (fungus - Sclerotium oryzae): Stem rot becomes most noticeable in rice fields during the latter stages of maturity. The disease occurs in circular to irregular areas in fields and causes premature death and lodging of the plants. The fungus attacks the rice plant near the water line usually during late tillering or early reproductive stages of growth. It first causes black, rectangular lesions with distinct angular borders on the leaf sheath. Later the lesions become larger, more diffuse, irregular in shape, and penetrate deep into the culm.

As rice approaches maturity, injury to the stems increases and reaches its peak at harvest. Weakened stalks break during this stage and plants lodge making harvest difficult. Plants infected early yield poorly. Ratoon cropping in many areas is impractical because of the high percentage of plants killed by the disease. Diagnosis is confirmed by obtaining an infected plant, splitting the base of the stem, and observing the presence of tiny, black sclerotia in internal stem tissues. Control measures include the following: crop rotation, use of early maturing varieties, fluctuating the flood water level, avoiding excessive rates of nitrogen, and rice stubble destruction. Some fungicides help to suppress this disease but are not highly effective.

Straighthead

Straighthead (physiological disorder): Straighthead is a physiological disorder that causes the entire head to be blank and remain upright at maturity. Straighthead generally occurs in spots scattered throughout a field and is most easily recognised near harvest when normal plants have downturned heads. The disease is frequently found on sandy loam soils but seldom on clay soils. Old cotton fields with arsenic residues can have a severe incidence of straighthead. Other, as yet unknown, soil factors are also involved in causing straighthead. Often it is found in fields where excessive non-decayed vegetation has been plowed under just before planting.

The disease is characterised by upright heads when the rice matures due to infertile seed. Hulls may be distorted into a crescent shape or "parrot beak". One or both hulls may be missing. Affected plants continue to grow, are a darker green, and often produce shoots from a lower portion of the plant. First year crops of rice grown on "new ground" are more likely to be affected. Some varieties are more tolerant than others. Control measures include planting resistant varieties and draining fields with a history of the disorder just prior to internode elongation.

DISEASES OF WHEAT

Common Root Rot

Common (dry land) Root Rot (fungus - Bipolaris sorokiniana): The fungus infects the sub-crown internode of young wheat seedlings. Root rotting can continue throughout the season and is most severe when new root formation is slowed. Later in the season, drought and warm temperatures favor root rot. Seed treatment and balanced fertility programs can reduce disease severity. Crop rotation may also reduce the disease. Common root rot is widely distributed in Texas.

Foot Rot

Foot Rot (fungi - Fusarium avenaccum, Fusarium spp.): Foot rot generally begins as an invasion of seedling roots. Foot rots are favored by cool temperatures. The fungus is more a secondary invader of roots initially infected by the common root rot fungus. The fungus surviving on crop residues and rotation may reduce disease severity

Glum Blotch

Glum Blotch (fungus - Septoria nordorum): Glume blotch occurs on the nodes, spikes, and glumes causing blackened areas. Stems are weakened and may bend or break just above the nodes. Seed may be shriveled, reducing yields and quality of grain. Seed treatment, rotation, deep burial of crop residue, foliar fungicides, and resistant varieties will reduce disease losses.

Leaf Blotch

Leaf Blotch (fungus - Septoria tritici): Lesions appear as pale green to yellow spots on the leaf. As cells in the spot are killed it turns brown. Lesions also appear on the head. Later, gray to black fruiting bodies are formed in the dead spot. Control with sanitation (early plowing) and crop rotation.

Leaf Rust

Leaf Rust (fungus - Puccinia recondita): Leaf rust occurs on either side of the leaf and on the leaf sheath as small, reddish-orange pustules (rust spores). In most years, leaf rust causes more damage in Texas than any other wheat disease. Growth is rapid between 59 - 72° F. It causes a reduction in the number and size of kernels. The disease reduces forage production in fields where it is utilised for grazing. New races of the rust fungus originate naturally and challenge wheat varieties. The leaf rust fungus of wheat does not attack oats or barley. Control with resistant varieties and foliar fungicide applications.

Loose Smut

Loose Smut (fungus - Ustilago tritici): Loose smut destroys the grain and all glume structures of the spike leaving only the central rachis. which head early. Infected plants, produce smut spores that are wind-borne to healthy plants at flowering time. Spores germinate and germ tubes penetrate the young wheat ovaries where the fungus remain dormant until those seed germinate. Use of seed treatment fungicides and disease-free seed will prevent this disease.

Powdery Mildew

Powdery Mildew (fungus - Erysiphe graminis f. sp. tritici): Powdery mildew is usuall found only on the leaves, but the fungus may attack all aboveground parts of the plant. It is noticeable first as small, irregular or circular, light gray spots on the upper surface of a leaf. The spots enlarge as the fungus grows and often may involve large parts of the leaf. As the spots age, the fungus on them takes on a flowery appearance which is due to the production of an enormous number of spores. Often the lower surface of the leaf beneath the diseased spots turns yellow and older parts of the spots turn brown. Affected leaves become deformed and crinkled and in severe cases they become brittle or they may die prematurely. As affected wheat approaches maturity, small, black fruiting structures (seen as black specks) may be scattered throughout the fungus growth on infected spots. This disease is not seed-borne but can be soil-borne. Close grazing to keep top growth reduced to a minimum will allow sunlight and air to keep the crown of the plant dried out, thus reducing the occurrence of powdery mildew. Rotations with non-host plants will also help reduce the soil-borne phase of this disease. Chemical control may be economical on high yield potential wheat.

Rhizoctonia Root Rot

Rhizoctonia Root Rot and Sharp Eyespot (fungus - Rhizoctonia solani): Many *Rhizoctonia* strains infect wheat roots and culms. Culm infections are sometimes observed in North Central and East Texas. These are referred to as sharp eyespot. Sharp eyespot begins as a gray colored lesion on the lower leaf sheath. Lesions may later turn straw colored. Acid, sandy and dry soils increase disease risk as do cool spring temperatures. Bare Patch of wheat is also caused by a particular strain of *Rhizoctonia solani*. This disease has not yet been identified in Texas.

Stem Rust

Stem Rust (fungus - Puccinia graminis f. sp. tritici): Stem rust is recognised by the elongated, ragged pustules it produces on stem, leaf, sheath, blade, chaff, beard and occasionally on young kernels. Fragments of epidermis adhere to the sides and ends of the pustules, giving a ragged appearance. The brick-red color and large elongated pustules distinguish it from leaf rust which has small round pustules and orange-red spores. Stem rust is far more devastating than leaf fust on susceptible varieties. Soft wheat varieties are generally more susceptible than hard red winter wheat varieties. Barley, rye, wild barley, and goatgrass are also susceptible

Stinking Smut or Bunt

Stinking Smut or Bunt (fungus - Tilletia foetida): Heads affected by stinking smut fungus have a distinct blue cast. At bloom time infected heads are more slender than healthy heads and do not put out pollen sacks. At maturity they appear plumper but lighter in weight than normal heads. The smut ball consists of a mass of foul-smelling, dark-brown powder (spores of the stinking smut fungus). In the field, smutted heads usually stand more nearly erect than healthy heads because of their lighter weight. In some varieties, it is necessary to crush the kernels to determine if heads are diseased. The offensive odor indicates the presence of heavy infections either in fields or in shipped grain. Infested wheat is usually condemned before shipment. While smut balls may be removed by cleaning and recleaning seed, spores will still be carried on kernels. Chemical seed treatment should be used to control spores on seed.

Stripe Rust

Stripe Rust (fungus - Puccinia striiformis): Stripe rust resembles leaf rust except the pustules develop along the leaf veins as long streaks. Mild winters followed by very cool springs and abundant rainfall permit this disease to increase.

Take All

Take All (fungus - Gaeumannomyces graminis var. tritici): Take-all is most obvious near heading on plants growing in moist soil. Diseased plants have basal internodes that are shiny black and have few tillers. Heads ripen prematurely and are bleached white and sterile. Plants may break free at the crown when pulled from soil. The fungus causing take-all persists in infected wheat stubble. Take-all is favored by alkaline, compacted, infertile (esp. nitrogen and phosphorus-deficient) and poorly drained soils. The disease has only been observed in Texas under irrigated conditions on the High Plains and Rolling Plains. Rotation may reduce disease incidence in fields where take-all has been a problem.

Tan Spot

Tan Spot (fungus - Pyrenophora trichostoma): Tan spots can occur on both leaf surfaces. Initial symptoms will be tan-brown flecks and the spots will enlarge into lens-shaped lesions. The fungus grows as a saprohpyte on crop residue. Because of this it has been observed to be more severe in no-till wheat. Cultural practices such as deep burial of straw and rotation will help reduce tan spot. Fungicides may be economical if the crop has a potential yield of forthy bushels or more per acre.

Barley Yellow Dwarf Virus

Barley Yellow Dwarf Virus: New growth of wheat plants infected in the seedling stage is chlorotic or yellowish in color. The entire plant will be severely dwarfed, tiller sparsely and produce few heads with little or no seed. Plants that become infected after the tillering stage are not dwarfed. The yellow dwarf virus can be carried from plant to plant by several species of grain aphids that overwinter on perennial grasses. No effective control measures are known other than utilising tolerant varieties

Soil-born Mosaic Virus

Soil-borne Mosaic (virus): Symptoms of soil-borne mosaic

appear early in the spring, usually when weather is cool and damp, but are rarely seen in fall or winter. Fields observed from a distance have irregular patches of light green or yellow plants. Symptoms on plant leaves range from mild green to yellow mottling and striping, giving the mosaic appearance. Stunting varies from moderate to severe and may be accomplished by rosetting. Symptom expression favors temperatures below 68° F. Symptoms gradually disappear before harvest time if normal temperature persists. A soil-borne fungus, *Polymyxa graminis*, which is a parasite of roots of many grass plants, is the vector of this disease. Virus particles are either inside spores of the fungus or are attached to the spores. The fungus invades the roots in the fall, when soils are cool and wet, carrying the virus particles with it. There is no practical way to rid the soil of the soil-borne mosaic virus. Rotation out of wheat to other crops will reduce losses.

Wheat Streak Mosaic Virus

Wheat Streak Mosaic Virus: The vector of this virus is a microscopic eight-legged, cigar-shaped, wheat curl mite, *Aceria tulipae*. This disease is more severe in the High Plains area of Texas. Although most severe infection occurs in the fall, the characteristic yellowish streaking and mottling of leaves usually are first observed after warmer spring weather. As plants approach maturity, the leaves turn brown and die. WSMV causes red streak on corn kernels. Control for this disease is possible by clean tillage around fields, early destruction of volunteer wheat, late planting after frost, and resistant varieties.

REFERENCES

Nordbo, E., Christensen, S., Kristensen, K., Walter, M., "Patch spraying of weed in cereal crops", *Aspects of Applied Biology*, 1994.

Nordmeier, H. & Dunker, M., "Variable weed densities and soil properties in a weedmapping concept for patchy weed control", *In:Precision Agriculture '99*, Odense, Denmark, Seccond European Conference on Precision Agriculture, 1999.

11

Diseases of Trees

ASH

Fraxinus spp.

Anthracnose (fungus - Gloeosporium sp.): Large areas of the leaf, especially along the edges and veins, turn brown. Premature defoliation will follow severe anthracnose infection during wet seasons. The disease may be confused with problems caused by weather adversities or other physiological problems. Spraying with a fungicide two or three times at 14 day intervals during humid periods will control the disease.

Leaf Spots (fungi - Cylindrosporium sp., Marssonina sp.): Among the most common foliage diseases of ash that occur virtually wherever ash is grown. Lesions appear early, hundreds may develop on a single leaflet. They are very small at first. Spray as recommended for anthracnose to control the disease.

Rust (fungus - Puccinia sparganioides): Swollen and distorted gall-like structures occur on leaves and twigs. The orange swellings are more common early in the growing season, particularly in the Gulf Coast area. Two stages of the rust fungus are known: one which occurs on ash trees, the other on grasses, Spartina sp. Spraying with fungicides at two to three week intervals during early spring will control the disease.

Cotton Root Rot (fungus - Phymatotrichum omnivorum): Ash is moderately susceptible to the cotton root rot fungus, particularly during the younger stages of growth.

Wood Rots (fungi - Poria sp., Fomes sp., Polyporus sp., etc.): Most of these fungi attack only weakened or wounded trees. Infection usually takes place through wounds caused by lawn mowers, pruning, or strong winds. Trees decline slowly for no apparent reason and the fungus slowly rots the wood. After the disease has progressed for some time, leathery, hard structures (conks) can be seen attached to the lower parts of the trunk. These are fruiting bodies of the fungus appearing as bracket growth during certain times of the year. Control is accomplished by avoiding mechanical wounds, treating exposed wounds with pruning paint, fertilising trees as needed and protecting from insects.

Powdery Mildew (fungus - Phyllactinia guttata): White powdery fungus growth on the leaves during the summer, then forming small black round fruiting structures in the late summer and fall. Damage is usually not extensive enough to warrant control.

Hairy Root (bacterium - Agrobacterium rhizogenes): A large number of very small roots develop either from the base of the stem or the larger roots. No control is known.

Cankers (fungi - Cytospora sp., Diplodia sp., Dothiorella sp., Nectria sp.): Several fungi cause branch and trunk cankers on ash. None of them are very common. Prune out infected branches. Maintain the trees in good condition by fertilising, watering and spraying for insects.

BOXELDER

Acer negundo

Verticillium wilt (fungus - Verticillium albo-atrium): This is one of the most common fungal diseases of plants attacking over 300 species. Boxelder is a member of the maple family and this disease is the most important killer of trees in that family. Symptoms include sudden wilting, usually of individual branches, followed by yellowing and finally browning of the foliage. Extensive internal streaking of the outer sapwood. Infected trees should be removed and burned. If a tree is to be saved for as long as possible, prune off all dead and dying branches. Cutting tools should be sterilised with 10 percent bleach between cuts. Inspect the tree periodically for dying branches.

Anthracnose (fungus - Gloeosporium sp.): Spots are indefinite, light brown, and may enlarge and run together causing the death of entire leaves. Leaves partially killed appear as if scorched. The disease is more common during prolonged rain periods when it may cause severe defoliation of trees. Several applications of fungicides may be required to control the disease during periods of high humidity.

Leaf Spot (fungi - Phyllosticta minima, Rhytisma punctatum, Cristulariella pyramidalis): These fungi attack many species of the maple family (*Acer*) including boxelder. The following fungi attack only boxelder: *Phyllosticta negundinis, Cylindrosporium negundinis, Septoria negundinis, and Cercospora negundinis.* Many leaf spots, one-fourth inch or more in diameter, some irregular, some with indefinite margins, occur on boxelder. Spots have brown margins and light colored centers. Some of the spots may develop black bodies, which are the fruiting structures of the fungi. Diseased, fallen leaves should be raked up and burned. Spraying the tree several times at two week intervals may be needed to keep the disease in check.

Powdery Mildew (fungus - Uncinula circinata): A white to grayish mold develops on the surface of the leaves. The disease is seldom serious on boxelder.

Heart Rot (fungus - Fomes sp.): The fungus enters the trunk generally through wounds or branch stubs resulting from pruning or mechanical damage. The rot is generally found on dead timber, but the fungus can get established on living trees and destroy the heartwood. New conks, or fruiting bodies of the fungus may appear annually, but they are perennial structures, putting on another spore-producing layer each year. Best control is achieved through prevention. Avoid mechanical injuries to trees, such as would result from hitting the trunk with lawn mowers and other garden instruments. Paint wound with a pruning compound; remove and destroy conks. Make pruning cuts flush with the main trunk and do not leave branch stubs.

Herbicide damage: Boxelder is highly susceptible to 2-4-D, causing distortion, dwarfing, and blighting of foliage.

CATALPA

Catlapa Bignonioides

Verticillium Wilt (fungus - Verticillium albo-atrum): This is the major disease of ornamental catalpas. Trees wilt suddenly, defoliate, and have a brownish discoloration of the outer sapwood. There is no control. Heavy fertilisation sometimes enables infected trees to put a new ring of summerwood and springwood outside the infected area, stopping the radial spread of the fungus. Some trees may then recover.

Leaf Spot (fungi - Cercospora sp., Phyllosticta sp.): Round to irregular brown to black spots appear during prolonged periods of damp weather. Center of spots may fall out, leaving a "shot hole" appearance. Collect and burn fallen leaves. Chemical control usually not necessary. Spray valuable trees as leaves unfurl, when half-grown, and again when full grown.

Powdery Mildew (fungus - Microsphaera alni, Phyllactinia corylea): White to gray powdery mold on both leaf surfaces, usually late in the season on younger leaves. Leaves may be deformed or curled. Spray as needed with a powdery mildew control fungicide.

Trunk Rots (fungi - Collybia velutipes, Polyporus versicolor, and Polyporus catalpae): These tree fungi are heartwood rotters of catalpa that enter through wounds. Avoid wounding and keep the tree in good condition by fertilising and watering.

CEDAR, JUNIPER, ARBORVITAE AND CYPRESS

***Cedrus* spp., *Juniperus* sp. *Thuja occidentalis,* and *Taxodium* spp.**

Twig Blights (fungi - Diplodia sp., Phomopsis sp., and Coryneum sp.): Needles, twigs and smaller branches turn light brown to reddish-brown; gradually die back from tips. Sometimes serious on seedlings and young trees in wet seasons. Tiny, brown to black dots appear later on infected parts. Prune and destroy blighted parts. Spray regularly.

Winter Injury (non-pathogenic): Injury evident in late winter and spring. Previous year's foliage is scorched, turns brown and dies back from tips of branches. Water plants during dry winters. Mulch during winter to conserve moisture and prevent freeze damage to roots.

Root Rot (fungus - Phytophthora sp.): Foliage wilts and fades to tan, yellow or light brown. Branches or tops die back. Grow in well-drained areas.

Rust (fungus - Gymnosporangium spp.): Greenish-brown to reddish-brown, corky, round to irregular galls on leaves and small branches. Masses of bright orange to brown, jelly-like spore horns form in wet spring weather. Prune out and destroy all infected areas.

Sooty Mold (fungus - Capnodium sp.): Unsightly, dark brown on black blotches coating the needles and branches. This is a fungus that usually grows on honeydew excretions made by insects (e.g., aphids, scales, white flies and others) or in flowing sap. The only damage caused by the fungus is through shading. Control insects.

Annosus Root Rot (fungus - Heterobasidion annosum): The greatest enemy of Eastern red cedar over much of its range. This fungus completely destroys the living bark and sapwood of the roots to the root collar, but seldom extends into the trunk. The fungus fruiting structures, irregular whitish conks, are found under the duff at the ground line. Most common in East Texas.

COTTONWOOD AND POPLAR

Populus spp.

Cytospora Canker (fungus - Cytospora chrysosperma): Primarily affects weak trees and occurs mainly on stems. Cankers first appear as slightly sunken areas on the smooth bark of branches and trunks. Cankers generally develop in an elliptical pattern and enlarge until stems are girdled and killed. Diseased part becomes discolored and sapwood is reddish-brown and watersoaked. During moist weather, reddish threads of fungus spores ooze from pimple-like fruiting bodies on cankered areas. Infected limbs should be removed, as the fungus moves down the stem and invades larger branches or even the trunk. Tree surgery to remove cankers in the main trunk may prolong the life of the tree. Sterilise pruning tools. Avoid wounding trees. Simon and Lombardy varieties are extremely susceptible - Rio Grande is resistant.

Dothichiza or Branch Canker (fungus - Dothichiza populea): This disease mainly attacks young planted and nursery trees. Dark, sunken, cracked cankers form on twigs, trunks and branches. Callus tissue may form in old cankers. Sapwood is discolored. Wounds are primary source of entry for the fungus and should be avoided. Lombardy is by far the most susceptible species. No effective control measures are known. Pruning of diseased parts does not help control this disease, in fact, it spreads it.

Septoria Canker (fungus - Septoria musiva or Mycosphaerella populorum): Cottonwood and Lombardy varieties are susceptible. Leaf infection precedes twig or stem infection. Canker is not easily distinguished from *Cytospora* canker. The canker on highly susceptible young stems appears sunken, with smooth bark and raised, irregularly concentric rings of unbroken bark. Some cankers are flat, others are swollen at the margins. Identification usually depends on finding little black fruiting structures (pycnidia). Remove infected limbs. Avoid wounding trees and sterilise pruning tools. Preventative fungicide sprays may control the leafspot stage.

Powdery Mildew (fungus - Uncinula salicis): A white fungus grows on both sides of the leaves. Damage is usually not serious enough to warrant control measures, but fungicides can be used on valuable trees.

Heart Rot (fungus - Fomes spp.): Primarily a disease of heartwood. Discoloration and soft rot of heartwood. Distinguishing feature is the shelf-like fruiting structures or conks formed on the main trunk. Trees with conks may live for several years. Remove only trees that are structurally unsound or completely dead. Avoid wounding healthy trees.

Wet Wood (bacterial): Occurs in Cottonwood and Lombardy Poplar. This disease is caused by a bacterial infection.

Rust (fungus - Melampsora medusae): Small, yellowish-orange pustules on lower surfaces of leaves. Pustules later turn dark brown or black. Generally, most severe during cool fall weather, but not severe enough to warrant control.

Fungal Leaf Spots (fungi - Cercospora populicola, Septoria spp., Marssonina populi): Irregular to round lesions, from brown to gray in color. Spots may coalesce and cause defoliation with severe infections. The most common leaf spot is caused by *Marssonina populi*. Large brown spots with darker brown margins cause premature defoliation. Twigs are also invaded and killed. Foliar fungicides should be used on a regular basis.

Crown Gall (bacterium - Agrobacterium tumefaciens): This organism has been reported to cause branch and root galls on cottonwood in Texas.

Cotton Root Rot (fungus - Phymatotrichum omnivorum): Cottonwood and Lombardy Poplar are moderately to highly susceptible.

DOGWOOD

***Cornus* spp.**

Sun Scald: Death within the first two years after transplanting is most commonly due to sun scald, lack of soil moisture or careless transfer of trees too large or without adequate root systems. Transplant small trees into partial shade during December of January. Container grown nursery trees are less subject to mortality. Wrap tree trunk loosely with paper. Water and fertilise properly.

Ascochyta Leaf Blight (fungus - Ascochyta cornicola): This fungus produces irregular leafspots varying in size, with gray to tan centers and prominent borders. The leaf may completely collapse, shrivel, and turn black.

Leaf Spot (fungus - Cercospora sp.): Spots of various sizes with dark brown to purplish borders and grayish centers. Not normally serious, but chemical control may be desirable if disease reappears annually.

Botrytis Petal Blight (fungus - Botrytis cinerea): This fungus disease affects foliage and green shoots as well as petals. This is a disease of wet spring weather that often follows frost damage. Bracts or "petals" rot in irregular brown patches. During very wet weather these rotting bracts are covered with a gray-brown, fuzzy mold that produces powdery spores. Spray early when flower buds start to open.

Cotton Root Rot (fungus - Phymatotrichum omnivorum): Dogwood is extremely susceptible to the Cotton Root Rot fungus, but also is not adapted to the root rot areas of Texas.

ELM

***Ulmus* spp.**

Black Leaf Spot (fungus - Gnomonia ulmea): Small, yellow spots appear first on upper surface of leaves, then gradually develop a shiny black appearance. Heavy spotting causes leaf yellowing and early defoliation in wet seasons. Usually defoliation does not occur much before normal leaf fall so control is not warranted. If trees have been affected seriously in previous seasons, fungicidal sprays applied when leaves are unfolding, when they reach full size, and again two weeks later will help prevent serious defoliation. Raking and burning fallen leaves will reduce inoculum for future infection.

Other Leaf Spots (fungi - Gloeosporium sp., Cercospora sp., Phyllosticta sp., and others): Dark, elongated spots develop on midribs, veins and margins of leaves, or spots of various shapes and colors may develop on any portion of leaf surface. Destroy fallen leaves and control as for black spot.

Wet wood or Slime Flux (bacterium - Erwinia nimipressuralis): Chronic bleeding of sap from crotches, wound or other weakened areas of trunk, with unsightly discoloration of bark in affected area. Sap frequently is sour smelling. Bleeding or fluxing is most pronounced during spring months or during wet weather. The problem results from fermentation processes of the causal bacteria creating pressures up to 60 pounds per square inch within the tree. Tapping directly into the trunk just below the affected area to provide an outlet for abnormal sap and gasses will relieve internal pressure and may aid in recovery. Drill a small hole (one-half inch diameter or less) directly below the bleeding site and slightly upward into the center of the trunk. Install a tight fitting drainpipe in the drilled hole making sure the end of pipe extends far enough outward so that sap does not fall on the tree.

Dutch Elm Disease (fungus - Ceratocystis ulmi): Symptoms may appear on one or more branches on any part of the tree in contrast to phloem necrosis where tops of infected trees show first abnormalities. Leaves on individual branches wilt and turn yellow; in some instances leaves wilt very rapidly, dry out, then fall while still green. Twig terminals of affected branches sometimes become curved to resemble a shepherd's crook. As a further diagnostic aid, twigs when cut across, show discoloration or browning of water-conducting tissues in the sapwood. Tree defoliation may occur rapidly or take place over an entire season. Likewise infected trees may die in a single season or live for several years. The disease is spread by elm bark beetles infested with the causal fungus.

Elm Leaf Scorch - (bacterium) - A rickettsialike bacterium has been associated with this condition. Vascular bundles are plugged to the point where water movement in the tissues is impaired. No control is known.

Powdery Mildew (fungi - Phyllactinia guttata, Uncinula macrospora, Microsphaera alni): Powdery whitish to gray growth on both sides of leaves. Affected leaves may be cupped, stunted, and show yellowing. The disease usually occurs so late in the growing season that chemical control is not necessary.

MAGNOLIA

***Magnolia* spp.**

Leaf Spots (fungi - Coniothyrium sp., Epicoccum sp., Glomerella sp., Phyllosticta sp., and Septoria sp.): Various sized leaf spots generally developing in late fall and early spring. Generally cause little damage.

Algal Leaf Spot (alga - Cephaleuros virescens): Forms a velvety coating with reddish-brown hair-like structures on the underside of the leaves. Occurs mainly in the Gulf Coast area. Causes little damage but treatment with fungicides may be required in some cases.

Leaf-Shedding: Magnolia will begin dropping leaves in November and continue until early spring. This is a natural phenomenon and the degree of leaf drop will depend upon genetic as well as physiological factors affecting the tree. Some trees may be completely denuded before new foliage comes out.

Wood Rots (fungi - Fomes sp., Polyporus sp., Daedalea ambiqua): Rot of heart wood. Colors range from white to black, may also be green or red or brown. Lines of infection are conspicuous. Conks may be formed on main trunk near soil line. Avoid wound. Infected portions may be removed by tree surgery.

Winter Injury: Some trees that harden off gradually in the fall can take temperatures as low as 0 degrees F and show no damage. Other individuals are hurt easily if temperatures commonly get below 15 degrees F. Killing of leaves and shoots can be extensive. Winter drought coupled with low temperatures can cause extensive die back or death.

MAPLE

Acer spp.

Anthracnose (fungus - Gloeosporium apocryptum): In rainy seasons this disease may be serious on silver maples and Japanese maples. Irregular, light to reddish-brown, or purplish-brown, dead areas occur on the leaves. Many spots occur along the veins. Areas often enlarge killing the entire leaf. Leaves partially killed appear as if scorched. Many infected leaves drop in late spring. Twigs may die back. Collect and destroy fallen leaves. Spraying with a recommended fungicide when buds begin to unfold in the spring and again in 10 to 20 days will help prevent infection.

Leaf Scorch (physiological): Light or dark brown areas along the leaf margins extending toward the mid vein. Foliage appears bronzed, dried and scorched. Causes: late spring frost; hot, drying summer winds; and drought. Water trees during summer droughts. Plant in areas protected from drying winter and summer winds.

Wilt (fungus - .Verticillium albo-atrum): Sudden wilting and drying of leaves on individual branches, particularly on one side of the tree. New leaves may be reduced in size and turn yellow. Infected trees may die slowly or suddenly. An olive-green discoloration may develop in the sapwood. Other fungi can also cause this discoloration. The disease does not spread rapidly from tree to tree. Infection takes place most commonly through the roots because the fungus lives in soil. Some infections take place above ground through wounds caused by insects or by pruning. Trees showing general wilting of the entire tree cannot be saved. Recently infected trees with only a few wilted branches may possibly be saved by adequate fertilisation and watering. It is thought that leaf growth is stimulated, in turn enabling the rapid formation of a thick layer of sapwood, which seals in the infected parts of the trees.

Crown Gall (bacterium - Agrobacterium tumefaciens): Rough, irregular, swollen galls at the base of the trunk or on the roots.

Ganoderma Rot (fungus - Ganoderma lucidum): Rapid decline and death of tree. Large, reddish, and varnish-like mushrooms form at the base of the infected tree. No control known for this type disease.

Powdery Mildew (fungus - Uncinula circinata): White fungus growth on leaves. Rarely serious enough to use control measures.

Cotton Root Rot (fungus - Phymatotrichum omnivorum): Silver maple is rated as highly susceptible to *Phymatotrichum omnivorum*).

MESQUITE

***Prospopis* spp.**

Powdery Mildew (fungus - Erysiphe taurica): Infected leaves are covered with a white powdery material. When disease is severe, infected leaves can be distorted. Benomyl can be used for the control of this organism, however, control is not generally necessary.

Spongy Yellow Heart Rot (fungus - Fomes everhartii): The fungus enters through broken limbs or through insect tunneling. In advance stages the interior of the tree will be a soft, spongy yellow or brown mass. Perennial cankers are bracket to hoof shaped. The upper surface of the fungus is dark brown to black while the underneath is a dull reddish brown. Control is not recommended.

Rust (fungi - Ravenelia arizonica and Ravenelia holwayii): Rust attacks the leaves causing a distortion of the leaf mid vein. It is most often observed on the first leaves in the spring. Damage is usually not significant and control is not required.

Leaf Spot (fungus - Cercospora prosopidis): Although the fungicide benomyl would control the leafspot, its use is generally not necessary due to the limited damage caused by the fungus.

MIMOSA

Albizzia julibrissin

Mimosa or Fusarium Wilt (fungus - Fusarium oxysporum f. sp. perniciosum): The leaves wilt, dry and shrivel, although they may remain green or yellowish for some time. Later the leaves fall and the branch dies. Sometimes only one side of a tree may be affected the first year of onset of symptoms. Suckers may sprout from the main trunk, but there is usually no recovery. Brown discoloration can be found in the sapwood of trunks and branches even before the leaves wilt. Discolored areas of cut stems may appear as complete rings. Dark rings are also present in the roots. Infection is through the roots. The fungus can be carried over in seed from diseased trees. Two wilt-resistant varieties are available - Charlotte, with light-colored flowers, and Tryon, with deeper red flowers.

Leaf Spot (fungus - Cercospora glauca): The only described leaf disease of mimosa in this country. Angular, small, dark brown spot develops as disease progresses, the center of the spot may become pale. Disease generally not serious enough to warrant treatment.

MULBERRY

***Morus* spp.**

Bacterial Blight (bacterium - Pseudomonas syringae pv. mori): Watersoaked spots appear on leaves and shoots have black stripes. The leaves at the twig tips wilt and dry up. Some control is obtainable on young trees by pruning dead shoots in autumn and spraying with approved fungicides.

Leaf Spots (fungi - Cercospora moricola, C. missouriensis, and Cercosporella spp.): The leaves of mulberry are spotted by these fungi in very rainy seasons. The *Cercosporella* fungus can cause defoliation of older trees. Valuable specimens should be sprayed with approved fungicide if leaf spots are serious.

Popcorn Disease (fungus - Ciboria carunculoides): This disease, known only in the southern states, is largely confined to the carpels of the fruit. It causes them to swell and remain greenish, and interferes with ripening. The disease is of little importance. It does not lessen the value of the tree as an ornamental.

False Mildew (fungus - Mycosphaerella mori): The foliage of mulberries growing in the southern states may suffer severely from attacks of this fungus. It appears in July as whitish, indefinite patches on the undersides of the leaves. Yellowish spores emerge from the stomata on the underside and spread out so as to form a white, cobweb-like coating; the general appearance is that of a powdery mildew. The asexual spores are colorless, each composed of several cells. The infected leaves fall to the ground, and the overwintering or ascocarpic stage matures in spring on these leaves. Gather and burn all fallen leaves in autumn. Spray with approved fungicide mixture as soon as the mold appears in July.

Cankers: Cankers on twigs and branches and die back of twigs may be caused by at least six fungi: *Cytospora* sp., *Dothiorella* sp., *Gibberella baccata, F. moricola, Nectria* sp., and *Stemphylium* sp. These can be distinguished only by microscopic or laboratory tests. Prune and burn dead branches. Keep trees in good vigor by watering and fertilising.

Powdery Mildews (fungi - Phyllactinia corylea and Uncinula geniculata): The lower leaf surface is covered by a white, powdery coating of these fungi. Valuable specimens can be protected by occasionally spraying with approved fungicide.

Mushroom Root Rot (fungus - Armillaria mellea): Attacks in its usual manner with hardwoods, in that while its rhizomorphs may occur generally on healthy root bark, pathogenic invasion and damage by *A. mellea* is almost always preceded by major root injury or other debilitation of the host.

Cotton Root Rot (fungus - Phymatotrichum omnivorum): White mulberry has been rated highly susceptible to cotton root rot.

Root Knot Nematodes (Meloidogyne sp.): These nematodes attack seedlings in the East and Southwest. The problem can be serious on young trees but has less effect on older trees.

Summer Scorch (physiological): Leaves on affected trees have chlorotic margins. This condition is thought to be caused by low soil moisture and high atmospheric temperatures.

OAK

***Quercus* spp.**

Anthracnose (fungus - Gloeosporium sp.): Anthracnose is more common on white oak group (Bur Oak, Over Cup Oak, White Oak and Post Oak) than on the other more common oaks. The disease is first observed in early summer, causing irregular, brown spots. These will most often be adjoining veins and midribs of the leaves. They may cover most of the leaf area. The blotches and spots are irregular, distortion of leaves,and papery texture are diagnostic characteristics. Late in the growing season dark pustules, formed by the fungus, can be seen on the veins and midribs of infected leaves.

The fungus develops at its optimum rate under warm, moist conditions during the spring months. Maximum growth occurs at 83 degrees F and a relative humidity of 97 percent.

The fungus overwinters on dead twigs. To control anthracnose, destroy the diseased foliage, remove all dead twigs from the tree, and spray with a copper fungicide. Make first application when leaves are half grown and repeat in two weeks. The fertility program should also be increased to offset the weakening effect of continued defoliation.

Oak Leaf Blister (fungus - Taphrina caerulescens): Leaf blister is one of the major foliage diseases of oak trees in Texas. It is particularly damaging on red oaks and also attacks members of the white oak group. Leaves infected by leaf blister have discrete round, bulges in the leaf tissue with some cupping and twisting. The fungus overwinters on the bud scales. Infection will occur if the weather is cool and rainy during leaf emergence. The initial flush of leaves during extended periods of dry weather and above normal spring temperature are the only ones subject to infection by this organism. A copper fungicide just prior to bud opening is effective.

Actinopelte Leafspot (fungus - Actinopelte dryina): This disease is most severe on members of the red oak group but white oaks are also infected. The spots are small (2 to 5 mm in diameter), round and reddish-brown in color. The disease is most severe in late summer and early fall. Severe defoliation can occur. Actinopelte leafspot is a primary leafspot but can be found associated with other fungi. Infected leaves should be destroyed and trees which are severely defoliated by the fungus should be fertilised slightly more than normal to stimulate growth. Check foliage for possible minor element deficiency. Damage is sometimes associated with low levels of iron. Newly planted trees (first 3 years) are more subject to attack by this disease than well established trees.

Spot Anthracnose (fungus - Elsinoie quercus): Infection by this fungus can cause small spots on the foliage of red oaks. The spots are only on the upper surface of the leaf and are blackish-brown. The centers of spots are lighter in color and are generally 1 mm in diameter. Symptoms are first visible in early June and increase in number until the middle of August. By that time the foliage is severely damaged and the tree is a pale yellow. Defoliation occurs when infection is severe. Sanitation is important in the prevention and control of this problem. Remove all diseased leaves. Spraying is seldom required. Trees which have been severely damaged by the disease should be given additional fertiliser to stimulate new growth.

Powdery Mildew (fungi - Microsphaera alni and Phyllactinia guttata): Powdery mildew occurs on all groups of oaks. Infected leaves have a faint indistinct spot on the upper leaf surface and a white to off-white powdery growth on the lower side. The fungus will most often be found along the veins and midribs of the leaf. It is normally observed in late fall. Infected leaves will be slightly disfigured in severe cases.

Powdery mildew is seldom a problem on large individual trees. Damage can occur in nurseries and on low sucker limbs but rarely is disease sufficient to require spraying. Limb removal improves air circulation and will help reduce the occurrence of this disease.

Rust (fungus - Cronartium strobilinum): C. strobilinum (Southern cone rust) infects pines in mid-winter. The telial stage is only produced on an evergreen oak. They are thus closely associated with live oaks.

Infected oak leaves have yellow to orange pustules on the lower sides of the leaves. Severe infection may rarely cause defoliation. This most often occurs in late spring and early summer. The disease cycle is broken if the alternate host is removed; however, the spores can drift long distances between the oaks and pines. No control is normally required.

Hypoxylon Canker (fungus - Hypoxylon atropunctatum and other Hypoxylon spp.): The disease is first evident as a dieback of one or more branches. The foliage of the diseased limbs turns yellow and dries. This dieback continues from branch to branch through the stem until eventually the tree dies. This may require 1 or more years depending upon the environment and amount of stress experienced by the tree. Near death or shortly after tree death the outer bark sloughs off and exposes large masses of brown, dusty one-celled spores (conidia). These spores are gone within a few weeks and a grayish surface is visible. This is covered with numerous black fruiting structures. Mature fruiting structures (perithecia) can forcibly discharge sexual spores (ascospores) for distances of 60 mm. They are then blown to surrounding trees where infection occurs again. Entry appears to be through injured surfaces on limbs or trunk. The fungus grows best at 86 degrees F but can grow at 50 and 100 degrees F.

Hypoxylon canker causes a dark brown discoloration of the sapwood. With age the infected wood is lighter in color and has black zones or patterns in the wood when observed in cross section.

Hypoxylon canker occurs primarily on trees which are or have been in stressed conditions. Trees which have been damaged by excessive fill soil are often attacked by this organism. It is also suspected to be a fungus that can invade on oak wilt-infected trees.

Control is achieved by maintaining the trees in a healthy condition. Avoid injury to the trunk and limbs and never apply fill soil around the trees. Chemical treatments would not be effective because the fungus is located within the tree.

Physalospora Twig Blight (fungus -Physalospora glandicola): Affected trees look like they have been attacked by cicada. Dead shoots with attached brown leaves are found on the tree. The fungus enters through twig wounds. It eventually kills the twigs and branches. Although it causes noticeable damage to the tree, it seldom causes economic losses.

The fungus overwinters in cankers on twigs. Infected limbs should be removed and burned or destroyed during the summer. Make cuts 6 inches below the diseased area. Fertilisation, watering and pruning should be practiced to encourage tree vigor.

Endothia Canker (fungi - Endothia parasitica, Endothia gyrosa): Post oaks and red oaks are particularly susceptible. It has been associated with pruning cuts or limb breakage. Removal of cankers by pruning and increased tree vigor helps reduce losses.

Canker Rots (fungi - Irpex mollis, Polyporus hispidus and Poria spiculosa): Infection by these organisms tend to cause circular cankers which have a depressed center rather than a conk or mushroom. The center of the canker is a brown punk. They will quite often form after a limb has been damaged or broken off.

To reduce canker rots, remove any dead limbs or limb stubs in a tree. Prevent any injury to trees that might serve as an infection site.

Burls (fungus - Phomopsis sp.): Large swellings on the limbs and trunk are caused by a species of *Phomopsis*. Burls can occur on both oaks and hickories. There are some swellings which occur on white oaks that aren't caused by a fungus but rather by the wood naturally growing over young buds and forming a gall or swelling.

True burls can be removed by pruning or in some cases by surgery. They do little or no damage to trees.

Oak Wilt (fungus - Certocystis fagacearum): This is an important disease of all oaks. Members of the white oak group die very slowly after infection occurs and can carry the fungus in the vascular system without showing any symptoms. Red oaks die rapidly after infection occurs. Symptoms show up on red oaks in early May as a bronsing of the leaves. On live oak, the leaf symptom is variable. The most common symptom is brown necrosis of the leaf veins. The remainder of the leaf may remain green or turn slightly yellow. Severe leaf drop occurs while the leaves are still green. Cuts made through the wood may show discoloration in the last annual ring. Symptom development usually begins on one limb or branch and in time spreads rapidly to the remainder of the tree.

The fungus may be carried from tree to tree by various insects and through root grafts. Sap feeding beetles are important in the short range spread. Red oaks which wilt in the late summer or early fall develop spore mats under their bark during the next spring. As the mats develop, the bark sloughs off or ruptures exposing the fungus. Insects are attracted to the mats. If the insects move from mats to healthy trees which have open wounds the fungus can then enter the healthy tree and move into the water conducting tissue.

Control is obtained by destroying diseased trees immediately. Remove or burn stumps. To prevent spread between trees, cut a ditch two to three feet deep around an infected tree. Vapam can also be used for this purpose. To apply the vapam, drill holes six to eight inches apart and 15 to 24 inches deep. The Vapam is diluted one part to four parts water and one cup of the mixture is applied to each hole. As soon as the hole is filled, it should be plugged with soil. Before using the Vapam treatment or making a ditch, the presence of Oak Wilt should be confirmed by a professional pathologist. This is a severe treatment and can result in plant damage if not done properly.

Oak Decline (fungus - Cephalosporium diospyri): Live oak decline is a weak vascular pathogen of trees in Texas. Plants affected by the fungus first show signs of thinning out in the top of the tree which when first noticed, twigs in the upper portion of the tree show signs of dieback. The dieback will generally increase each year. As the dieback reaches the larger limbs which comprise the main canopy, sucker growth becomes evident on the main scaffold limbs. As the disease continues to progress, only the main scaffold limbs remain alive yet they also eventually die. This may take from five to ten years on Live Oaks, but less on other oaks after the first visual symptoms are observed.

Root grafting is suspected as one means of spreading the fungus. Recently, bark beetles have been associated with the spread of Oak Decline. The fungus has been isolated from the body of the suspect insects. If insects are involved this would explain the spread of the disease across areas too great for root grafting.

Laboratory isolations are not always satisfactory; this can lead to considerable difficulty in making any type of a reliable recommendation for control. For the most part, recommendations are based on positive identification of the problem. Mechanical injury to the roots can cause symptoms easily confused with disease caused by fungi. In newly built areas, the damage most often observed is from root injury.

Step to Reduce Live Oak Decline Losses:

1. Identify the problem.
2. Reduce the stress conditions that weaken trees.
3. Optimise fertiliser and water applications to tree.
4. Remove dead limbs on tree.
5. Do not import wood from Oak Decline infected areas as it could be a source of the disease spread into a previously disease free area.
6. Use trees not known to be a host of Oak Decline.

DISEASES OF PALM

False Smut symptoms on *Phoenix dactylifera*

False Smut (fungus - Graphiola phoenicis): Small, black, wart-like bodies are seen, often with yellow filaments protruding from the middle of the raised spots. Both sides of the leaf are affected. A high incidence of the fungus on leaves can lead to premature death of older leaves. This is a common disease of date palms in south Texas, but chemical control is seldom required. Fungicides will protect against the disease.

Leafspots symptoms on *Sabal* palm

Leafspots (fungi - including Bipolaris spp., Exscrohilum spp., Colletotrichum spp., Annellophora spp.): The spots are usually small, brown to black, and occur mainly on older leaves.

Small, black fruiting bodies may be seen in dead tissue. The presence of a few fungal leafspots does not usually indicate a serious disease problem. They may occur on tissue that has been weakened as a result of an environmental stress such as poor drainage. Alleviating the stress will end the problem. Chemical control is seldom required. Sometimes copper-containing fungicides can cause leafspots that resemble those caused by fungi. Potassium deficiency can also cause leafspots. Fungicides can be used to protect against leaf-infecting fungi when there is high disease pressure (i.e. sustained humid or wet conditions).

Lethal Yellowing Disease symptoms on Palm

Lethal Yellowing Disease (mycoplasma): In the Rio Grande Valley, the Canary Island date palm (*Phoenix canariensis*) and the true date palm (*Phoenix dactylifera*) are affected, although other species of palms found in other areas are also susceptible. The disease is first noticed when lower leaves turn yellow, then brown and droop. The wilt spreads to upper leaves. As all the leaves collapse, the appearance resembles a "closed umbrella." Eventually, all the dead leaves drop and only the trunk remains. This process takes about 4 months following observation of initial symptoms. The disease has been controlled in other areas by repeated injections into the trunk with an antibiotic (oxytetracycline). However, the treatment is not a cure; it only alleviates symptoms and must be repeated every three to four months. Species adapted to Texas that are resistant include: Mexican fan palm (*Washingtonia robusta*), California fan palm (*Washingtonia filifera*), pygmy date palm (*Phoenix roebelenii*), Texas palm (*Sabal texana*), and Queen palm (*Syagrus romanzoffiana*).

Rachis Blight symptoms on Palm

Rachis Blight (fungus - Serenomyces californica): This disease has been found on the California fan palm (*Washingtonia filifera*). Portions of the rachis, running longitudinally, have a brown discoloration. The inside of the rachis is also discolored. As the disease progresses, fruiting bodies of the fungus are formed within the rachis. As they mature, the rachis surface is broken, revealing masses of dry, rust-colored spores. Distinct portions of the leaf are killed; the dead area is associated with where its vascular system originates from the diseased area of the rachis. There is no treatment known, although removal of infected foliage may prevent spread to other foliage.

REDBUD

***Cercis* spp.**

Canker (fungus - Botryosphaeria dothidea): The most destructive disease of redbud. Cankers begin as small sunken areas and increase slowly in size. The bark in the center blackens and cracks along the edges. The canker girdles the stem and the leaves above wilt and die. Prune and burn branches with visible cankers.

Leaf Spot (fungus - Mycosphaerella ceridicola with a Cercospora imperfect stage): The leaf spots are brown, circular to angular, from one-eighth to one-fourth inch in diameter, often with a halo. Premature defoliation may occur. Apply a fungicide labeled for general use on trees and shrubs, at 10 day intervals when spots first appear and during prolonged periods of wet weather.

Cotton Root Rot (fungus - Phymatotrichum omnivorum): Redbud is rated as highly susceptible to the cotton root rot fungus.

SYCAMORE

***Platanus* spp.**

Wilt (fungus - Cephalosporium diospyri): This fungus has been found repeatedly in large sycamores. The fungus, when introduced into seedlings, produces symptoms including sudden wilting and browning of leaves, sudden blighting of twigs forming "shepherd's crooks", yellow discoloration, and defoliation. How widespread this fungus is in Texas sycamores is not known.

Botryodiplodia Canker (fungus - Botryodiplodia theobromae): This fungus is known generally as a weak or facultative parasite with a wide host range. However, in Texas it has caused a rapid death of sycamores. The fungus produces cankers on branches and the main stem. It is favored by high temperatures and drought stress conditions. Trees weakened by the Cephalosporium wilt fungus are more vulnerable to attack. Broken terminals in twigs are the best places for the fungus to enter. It has been a common problem in sycamores. Prune out dead and dying branches below the cankers. Sterilise pruning tools with 10 percent bleach solution between cuts. Spraying for anthracnose with benomyl will also help to control *Botryodiplodia*.

Sycamore Anthracnose (fungus - Gnomonia plantani): Sometimes called blight and scorch. A single attack seldom causes harm but if the tree is infected several years in succession it will weaken a tree, making it susceptible to borer attack and winter injury. The first symptoms, sudden browning of single leaves or clusters, may be confused with late frost injury. Later dead areas appear along or between the veins, usually starting at the leaf edge. Leaves fall prematurely when heavily infected and trees often remain bare until late summer, when new leaves form. Infection of small twigs causes sunken areas, called cankers, and slightly raised margins. When the canker completely girdles the twig, killing it, this is called the shoot blight stage.

The fungus can overwinter in fallen infected leaves and in twig cankers. In southern Texas it can pass the winter in the spore stage on dormant buds. Severity of attack depends on weather conditions during the two week period following leaf emergence. Frequent rains and cool temperatures favor rapid spread. Below 55 degrees F. injury will be severe and above 60 degrees F, little or no injury.

Control: Prune out dead twigs in fall. Burn all dead twigs and fallen leaves. Spray with recommended fungicide when leaves unfurl, when leaves are half-grown, and when leaves are fully grown.

Trees repeatedly attacked should be well fertilised in spring to increase their vigor.

Leaf Spot (fungi - Mycosphaerella platanifolia, Phyllosticta plantani, and Septoria platanifolia): Several fungi cause disease of minor importance that can be controlled by the spray schedule suggested for control of anthracnose. *Phloeospora multimaculans*:known mainly in Texas. Irregular, dark brown to purple spots one-eighth to one-fourth inch in diameter on the upper leaf surface. Often with a brown center. Spots are brown with darker border on lower surface. Spots may coalesce to produce a dirty-brown colored leaf. Defoliation may occur.

Powdery Mildew (fungi - Microsphaera alni and Phyllactinia guttata): Makes appearance mainly in late summer. May be present on older growth but usually most severe on new growth, which may be distorted, stunted, and covered with a white growth and white spores.

Canker-Stain Disease (fungi - Ceratocystis fimbriata f. sp. plantani): It can be lethal to sycamore, but is much more important as a killing disease of London plane (*Platanus acerifolia*): Since it is spread almost entirely by man, through pruning, it is essentially a shade tree disease. Leaves are dwarfed and sparse in part or all of the tree top. The staining cankers occur on trunks or branches. The first symptom on the yellow or green bark is a brown to black lens-shaped discoloration. Cankers may become 20 to 40 inches long in one year, but usually only two inches wide. Cankers widen each year, and often coalesce, girdling the tree or branch. Older cankers shed their darkened, dead bark exposing the wood, which dries, cracks, and turns black. The reddish-brown or bluish-black discoloration of the wood, in cross section behind the cankers, is the most distinctive symptom. Stain patterns are radial, generally reaching the pith. The fungus sporulates abundantly on newly-killed wood in wet weather from May until October. It produces two kinds of asexual spores; one, long and clear; the other, short and brownish. It also produces a long-necked, pear-shaped, sexual fruiting structure. Avoid injuring the tree. Prune out dead limbs during winter. Be sure to dip the pruning tool in 10 percent bleach solution before each cut to avoid spreading the fungus. Treat cankers with protective paint.

Sooty Blotch (fungus - Gloeodes pomigena): Sometimes forms on shoots of sycamore. The dark surface mycelium can usually be rubbed off with the fingers.

Mistletoe (parasitic plant - Phoradendron serotinum): A seed-producing higher plant that parasitises sycamore in the south.

Shoestring Root Rot (fungus - Armillaria mellea): Two forms of the fungus can usually be found under the bark at ground level. These are black "shoestring like" rhizomorphs and a white fan-shaped fungus growth.

Trunk Rots (fungi - Hydnum erinaceus, Fomes sp., Ganoderma sp.): Heart rot fungi can hollow out the entire central cylinder of a tree. The tree declines in general and the presence of the fungus is known when it fruits on the side of the tree. The annual fruiting structure is a white, rounded, spongy mass with long, slender white teeth on the bottom. There is no control - only prevention, by avoiding wounding the tree.

WILLOW

***Salix* spp.**

Bacterial Twig Blight (bacterium - Pseudomonas saliciperda):

Leaves turn brown and wilt and blighted branches die back for several inches. Brown streaks can be seen in sections of the wood. Bacterium overwinters in the cankers, so young leaves are infected as soon as they unfold. The damage can be confused with frost injury. Prune out infected twigs and spray in early spring with an approved fixed copper fungicide.

Crown Gall (bacterium - Agrobacterium tumefaciens): Mainly a nursery disease. Large, rough, woody swellings or galls on the lower part of the stem and crown of the plant. Infected plants may be deformed, stunted or even killed. Weeping willow is susceptible. No practical control is known for this disease.

Cytospora Canker (fungus - Cytospora sp.): Affects willows the same as poplars. Discolored, sunken, often sharply defined areas develop on twigs, branches or trunk. Cankers enlarge and gradually girdle infected parts causing death to portions beyond. Weeping willow is susceptible, but rarely occurs on black willow. Prune out and destroy dead and cankered parts. Spray with an approved fixed copper fungicide.

Leaf Spots (fungi - Cercospora sp., Gloeosporium sp.): Small to large, round to irregular spots of various colors on leaves. Leaves may wither and drop early. Begin spraying when buds begin to swell in the spring.

Rust (fungus - Melampsora sp.): Lemon-yellow spots on lower leaf surface. Later in the season the pustules are dark colored. The disease may be severe enough to cause leaf drop. Although rust infections are not considered serious, they may result in heavy defoliation of young trees.

Powdery Mildew (fungi - Phyllactinia guttata and Uncinula salicis): White powdery growth on leaf surfaces may become heavy late in the season, especially on tender leaves of sprouts.

Tarspot (fungus - Rhytisma salicinum): Spots are very thick, jet black, discrete and about one-fourth inch in diameter. It looks like a drop of tar on the leaf. Rake up an burn dead leaves, as the fungus overwinters on them. Spray early in April with a fungicide.

WOOD ROTS ON SHADE TREES

Wood decay is common on many species of shade trees. Wood decay fungi usually enter the tree through wounds. Most fungi grow rather slowly showing no evidence of being present until extensive areas are rotted. Most will first invade the non-functional heartwood. After decay is well advanced in the heartwood, they may invade the living sapwood.

When the fungus has developed extensively, fungal fruiting structures, commonly called shelf fungi or conks, may appear on the trunk. Some fungi may form mushroom-type structures at the base of the trunk or form on major roots, rather than on the trunk. Where extensive wood rot has already occurred, the rotted wood should be completely removed. The cavity then should be sprayed with a fungicide and filled by a reliable and experienced tree surgeon.

Improper pruning or failure to remove wind damaged branches is a common problem that results in fungal infection. Do not leave branch stubs. Cuts should be made even against the trunk so that callous tissue can grow over the wound. Cuts should be made so that the scarred area will never collect and hold water.

REFERENCES

Browne, F. G. "Pests and Diseases of Forest Plantation Trees", *An Annotated List of the Principal Species Occurring in the British Commonwealth*, Oxford, England: Clarendon Press, 1968.

Nyvall, Robert F., "Field Crop Diseases Handbook", Ames, IA: Iowa State University Press, 1999.

12

Methods and Materials to Cure Plant Disease

CULTURAL PRACTICES FOR REDUCING CROP DISEASES

Certain cultural practices are invaluable in reducing plant disease losses. A control program is enhanced whenever one can utilise as many methods of control as possible. New strains of an organism may develop that will attack resistant varieties or become tolerant to certain pesticides when these practices are used alone. Combining practices reduces the risk of failure.

Rotation with unrelated crops is probably the most utilised cultural practice for disease control. This helps keep populations of pathogens from building up to damaging numbers. One should not expect rotation to eliminate disease development, but it certainly aids in reducing damage from most diseases.

Fertiliser usage may have some bearing on development of certain diseases. It differs with each crop and each disease but, in general, nitrogen out of balance with other nutrients enhances foliage disease development and predisposes some plants to other diseases. Potash, on the other hand, helps reduce disease development when it is in balance with other elements.

Deep burial of crop residue helps control certain diseases by placing the organism contained in the residue at a depth where there is an oxygen deficiency. This reduces the population of the disease-causing organism and permits the crop to escape much of the damage.

Planting on a raised bed is helpful in preventing certain diseases such as Southern blight and certain of the wilt diseases. This practice is advisable when growing leguminous crops such as peanuts, soybeans, and guar, and when growing vegetable crops in tight, poorly drained soils.

Burning of crop residue has been discouraged because of destruction of valuable organic matter and creation of an air pollution problem. The fact remains, however, that it is a highly effective means of eradicating some disease-causing organisms associated with crop residue.

Time of seeding has an important bearing on disease prevention in many cases. Delayed planting of wheat will help escape the chances of wheat streak mosaic virus. Early spring planting of cotton may effectively help escape cotton root rot.

Removal of undesirable plants that might serve as a host reservoir for virus diseases that attack cultivated crops aid in preventing infection. Infected rhizomes of Johnsongrass are the primary overwintering host for the maize dwarf mosaic virus that attacks grain sorghum, forage sorghum, and corn. Wild Solanaceae weeds, such as jimsonweed, horsenettel and silverleaf nightshade harbor viruses that attack potato and tomato.

Volunteer plants from a harvested crop are often means of carrying a disease organism from one crop season to the next. Rusts of cereal crops and spinach constitute an example of this type disease dissemination.

Roguing (removal) of diseased plants as they appear is often an effective method in helping reduce the spread of a destructive disease. Virus diseases of stone fruits and bacterial wilt of cucurbits are examples where roguing is worthy of consideration.

These cultural control practices have been found to be economically feasible in reducing disease losses. Growers should properly identify the diseases that limit production and then use a variety of controls in combination.

DIAGNOSIS OF PLANT DISEASE AND NEMATODE PROBLEMS

Correct diagnosis is the first step in controlling a condition caused by disease organisms or nematodes. Since fungi, bacteria, nematodes, mycoplasma and viruses are microscopic in size, it is necessary to follow certain procedures to insure proper identification. The following series of steps will aid in taking a more direct approach.

— Carefully observe all symptoms associated with a condition. Formulate a description in your mind or on paper and check to see if the statement is true without exception.
— Compare plants to see if they are similarly affected in all parts of the field or yard.
— See if non-related plants are similarly affected. Most disease-causing organisms are host specific and do not affect a large number of plant types. If a similar leaf spot or burn is observed on different plant types, then one might expect a drift of toxic substances. On the other hand, certain diseases like cotton root rot might affect a number of plants, but you could rule out corn or other grasses, which are not susceptible.
— Get as much information as possible to help diagnose the problem. County Extension agents have a number of publications that will be helpful.
— If it is not possible to correctly identify the problem at this point, one should select a representative specimen for observation by the county Extension agent.

Loss estimates are necessary to properly evaluate damage caused by plant diseases. Development of this type of information is needed to determine what course of control action is needed and how much money can be spent on programs to reduce losses. Damage caused by disease organisms is often attributed to factors such as weather conditions instead of the disease itself. This natural tendency occurs because disease producing organisms are microscopic in size and not visible to the unaided eye.

It is rare indeed to be able to calculate the actual loss due to diseases and for this reason estimates are necessary. For example, a field of grain may be uniformly infected with an organism such as rust. This means that an unaffected area is not available for yield comparison. It is possible for one to use chemical control to protect an area but even then control may not be complete or the fungicides may alter plant growth.

Fungicide Additives

It is sometimes desirable to add certain materials to fungicides to improve final results. Additives or adjuvants can be placed in one of the following categories:

— *Diluents* —any liquid or solid material used to dilute active ingredient and make it less concentrated.

— *Surfactants*—materials added to a fungicide formulation to ensure that powders are easily wettable and well dispersed in the spray tank.

— *Spreader-stickers*—materials used to increase the spreading of spray droplets on plant surfaces and render the residue more tenacious.

— *Modifers*—compounds added to pesticide formulations to enhance biological activity or reduce plant damage.

Fungicide additives when needed are usually added to the product during formulation. This ensures product stability and performance. In some cases manufacturers will suggest that certain products be added at the time of use. If a label does not specify the use of an additive, one should realise that he assumes the risk of undesirable results if any should occur. Certain fungicides become toxic to plants when oils or spreader-stickers are added. The safest approach is to abstain from the use of additives unless the label so specifies.

Seed Treatment

Controlling a disease may often be achieved by a single procedure such as chemical treatment. Numerous plant disease organisms live over from year to year because spores of the pathogen are carried on or in the seed. When such contaminated seed is planted, the seed-borne microorganisms revive, start to grow and become active parasites, destroying the seed or attacking the seedling as it grows. One purpose of seed treatments by the use of chemical protectants is to destroy seed-borne microorganisms that cause seedling blights and seed decay. Many microscopic molds and bacteria live in the soil and many of these are pathogenic. Chemical seed treatment is used to provide a protective zone around the seed which soil-borne microorganisms cannot penetrate, thus protecting the developing tender seedling against attack until it is well established and has acquired the capability of outgrowing attacks from soil-borne pathogens.

Some plant disease-inciting fungi are carried within the seed. As the seed sprouts, the fungus grows inside the developing seedling, resulting in disease symptoms in the mature plant. Loose smut of wheat and barley is caused by a fungus of this type. Systemic seed treatment fungicides are necessary to control this type of fungus.

There is no one seed treatment method or material that may be recommended as a universal preventative for all of the pathogens that attack seed and seedlings. Certain liquid treatment materials recommended for small grains would be ineffective when used on seed such as beans and peas. Some of the seed treatment chemicals are poisonous to animals and humans and great care must be exercised when handling these products. Even though some seed treatment chemicals may not actually be poisonous, most of them like house dust, can be irritating to the eyes and nose. Chemically treated seed should never be used for livestock feed or human consumption. Seed treatment is a useful and indispensable tool in our defense against plant disease. Various methods and compounds for seed treatment are discussed in this publication.

Fungicide Adjuvants

It is sometimes desirable to add certain materials (adjuvants) to fungicide spray solutions to improve final results or to resolve problems with application. Adjuvants have been developed to improve spray dispersion, reduce volatility, spray drift, and foaming, improve plant penetration, buffer spray solution, plus other functions. The most commonly used adjuvants are non-ionic surfactants. Surfactants reduce surface tension of spray droplets which allows better plant coverage. Without surfactants, droplets roll off hard to wet plant surfaces.

Some fungicide formulations contain compounds that act as surfactants. In these cases manufacturers may not suggest a surfactant be added to the spray mixture. In other cases, and in particular with wettable powders, labels may emphasize the need for a surfactant to enhance effectiveness. If a chemical label does not specify the use of an adjuvant, one should realise that he assumes the risk of undesirable results if any should occur. The safest approach is not to use adjuvants unless the label so specifies.

Non-chemical Control of Plant Diseases

Plants are considered diseased when they do not grow and develop normally. The cause of such a condition may result from infection by disease-causing organisms or from environmental factors. Organisms known to cause disease are bacteria, fungi, nematodes, viruses, phytoplasmas and parasitic seed plants. Examples of environmental disorders are air pollution, poor soil, excessive heat, low or excessive nutrition, and drought.

When organisms cause disease, it is usually due to a parasitic relationship with a host plant. The parasite, or causal organism, is called a pathogen. The interaction between host and pathogen results in disease. This discussion is limited to diseases where a pathogen is involved. Most plant problems arising from environmental conditions are covered under such topics as fertilisation, water management and other cultural procedures.

Prevention is the best approach to plant disease control when using either chemical or non-chemical (organic) methods. Prevention may involve suppressing the disease agent or avoiding the disease. Utilizing as many disease-preventative practices as possible will ensure the best possible control. Many of the following practices aid in controlling many different disease problems that arise.

Suppressing the disease agent

Rotation

Vegetables in the same family group are more likely to be susceptible to the same soil-borne diseases. Cantaloupes and watermelons, for example, have common diseases. If they follow each other in a rotation, a disease organism may be limited on the first crop but sustain enough carry-over in the soil to cause severe loss on the following crop. Vegetables from different family groups should be rotated because they are usually not susceptible to the same disease organisms. The groups listed below should be rotated so a selection from one group is not planted in the same location more than once every 2 to 3 years. Limited garden space may prevent an ideal rotation system.

Organic matter

Organic matter increases the number and kind of microorganisms in the soil. Many of these microorganisms compete with disease agents for nourishment. In most cases, the best organic matter is obtained by turning under a green cover crop, such as a small grain (wheat, oats, barley, cereal rye) or a legume.

Resistant varieties

Agricultural scientists have made great strides in developing disease-resistant varieties. Resistance, however, is a relative term and does not indicate immunity. For example, a tomato plant resistant to Fusarium wilt may develop wilt if stressed but not become diseased to the degree that a susceptible variety might. Resistance could be expressed as slower disease development which allows resistant plants to produce an acceptable yield before or without losing vigor to disease attack.

Sanitation

Any crop residue destruction practice that reduces the disease agent's ability to reproduce or overseason could be included under sanitation. Examples are raking and burning diseased leaves and disposing of infected fruit. Root knot nematode-infested plants should be dug and as much of the root system as possible removed from the garden site. Root knot nematodes are harder to kill when protected by root tissue. Diseased leaves, fruits, and other plant parts should not be piled near the garden. Fungi, for instance, often produce thousands of spores in the reproductive process, and these spores can be wind-blown great distances. Diseased plant tissue should be buried, burned, or disposed of in some other way that prevents the dispersal of the disease agent. In most cases, it is best not to put diseased plant parts in a compost pile. This is especially true for nematode-infected plants. Although many disease organisms are destroyed by heat during composting, 100 percent kill is not normally obtained. Contaminated compost will only serve to spread disease organisms as the compost is being used. Compost for the garden could be made with grass clippings or shrubs and tree leaves, as the disease organisms on grass and leaves are not likely to affect garden plants.

Heat treatment

Heating small amounts of soil is practical for potted plants and for growing seedling transplants. The soil should be moist but not saturated. Heat the soil at 180°F for approximately 3 hours in a standard oven. An average-sised Irish potato placed in the middle of the soil can act as an indicator. When it is cooked, the soil should be sterile. Microwave ovens have also been used for sterilising small quantities. Using high power, heat the soil for 3 minutes. Do not overheat using either system. Excessive heating will release certain elements in the soil to toxic levels. A new heating approach called solar pasteurisation or solarisation was developed in Israel. It simply consists of covering well tilled, moist soil with clear plastic, sealing the edges, and leaving it for several weeks. Black plastic should not be used because soil heating will not be as great as heating under clear plastic. The best results are with fungi, principally those causing root rots. Certain weeds and nematodes are also controlled but not on a consistent basis. Solar pasteurisation will work best during the hottest months. August should be ideal for home garden sites. For best results leave soil covered 30 days.

Fallowing

Leaving land idle and clean through the growing season will reduce disease agents in the soil. Fallowing is especially helpful if done in the summer months when soil temperatures are high. Frequent plowing will keep the soil dry and free of plant growth and expose soil-borne disease organisms, such as nematodes, to killing heat and excessive drying. Other benefits of fallowing are weed and insect control. This practice is most efficient if rainfall is low and temperatures are high.

Weed control

Weeds harbor insects and serve as hosts for many virus diseases. For most viruses to survive they must remain in a living organism whether it be a host plant or insect. Destroying weeds in and around the garden may eliminate potential overwintering host plants.

Avoiding Disease

Change planting date

Some diseases can be controlled by changing the planting date. Spring-planted squash usually escapes mosaic virus, whereas fall-planted squash does not. Warm-season vegetables, like peas and okra, should be planted after soils warm sufficiently to avoid seedling disease.

Obtain disease-free planting stock

It is advisable to check transplants such as cabbage, tomatoes, peppers, and others for root knot nematode galls. Certified Irish seed potatoes are less likely to harbor a disease such as black leg. Some disease agents are seed transmitted; thus, only the best seed should be planted.

Cultural practices

— Select the best site. Choose a site with deep, well drained soil.

— Plant on raised beds. Raised beds improve drainage. They also warm up faster than level soil which aids in faster emergence. Slow emergence increases chances of seedling disease.

— Use proper plant spacing. Crowded plants reduce air circulation which enhances leaf spot diseases. Septoria leaf spot on tomato is an example. However, in Far West Texas, with low humidity and few leaf spot diseases, crowding tomato plants may reduce losses to curly top virus by shading the plants and making them less attractive to the insect vector.

— *Overplant.* Overplanting where possible will ensure enough production if some plants or fruit are lost to disease.

— *Proper fertilisation.* Plants receiving all needed nutrients may resist some of the less vigorous disease- causing organisms.

— *Do not overwater.* Excessively wet conditions increase soil-borne fungal diseases like root rots and wilts.

— *Avoid overhead irrigation.* Leaf spot diseases develop rapidly when leaves are moist.

— *Do not prune roots..* Roots cut while cultivating reduce plant vigor and leave openings for root rot and wilt fungi to enter.

Root knot nematode control

No doubt many gardens have become infested by planting contaminated transplants or by bringing in topsoil harboring root knot nematodes. Taking steps to prevent this problem is just as important as implementing the steps to control this once it has become a problem. Where soils are already infested and a garden is first put into production, nematode levels are generally low. After several years of gardening and growing susceptible plants, the nematode population increases to the point that damage becomes unacceptable. Control practices that can be used are summer fallowing, rotation, adding organic matter, planting trap crops, removing diseased plants, and using resistant varieties. All these control measures are designed to reduce the nematode population in the soil. The benefits of summer fallowing have already been discussed. Removing all potential hosts and keeping the soil dry will reduce the nematode population, as the nematode requires a moist environment for survival. Additional years of fallowing will further reduce the nematode population.

Rotation with non-hosts or poor hosts of the root knot nematode is also another means of reducing the population. Sweet corn is a poor host and is good to use in a rotation, especially in an area where root knot has done severe damage. Onions, garlic, asparagus, and shallots are also poor hosts. Cool-season crops such as cabbage, Irish potatoes, greens (turnips), radishes, and broccoli are less likely to suffer yield loss from root knot nematodes. Even though these are susceptible plants, they grow best in the cooler time of the year which is not favorable for root knot nematode development.

High soil organic matter alone will not ensure root knot nematode control. The higher the organic matter, however, the better the chance that antagonistic organisms will develop. There are soil fungi that trap nematodes and utilise them as a food source. Some organic matter tends to work better than others. Turning under a green manure crop such as small grains or legumes several weeks before planting has shown to be the best. Additional nitrogen may be necessary for adequate crop production because decay organisms in the soil will utilise available nitrogen as they break down the green manure crop.

Sometimes people resort to using home remedies to control nematodes such as planting marigolds or mixing sugar or lye into the soil. Of these three, only French marigolds are effective in controlling nematode populations, and their effectiveness has limitations and is often questionable.

Some people think that marigolds secrete a toxic substance into the soil that kills nematodes and that planting a few marigolds around annual plants in infested soil will prevent infection. This is not true. Marigolds primarily act as a trap crop. Nematodes are able to enter their roots but are unable to complete their life cycle. The trapped nematodes die without reproducing.

The type of marigold is also important. French marigolds, *Tagetes patula*, are more effective in controlling root knot nematodes than the African marigold, *Tagetes erecta*, which is also referred to as the American, Big, or Aztec marigold. To be effective, marigolds must be planted as a solid crop and grown for 90 to 120 days to effectively reduce the nematode population sufficiently to grow annual plants without treatment. Marigolds should be planted in rows no further than seven inches between each plant, so that the roots penetrate the entire soil mass to trap as many nematodes as possible. If marigolds are planted close together, they form a dense canopy which helps retard the development of weeds and grasses. Many weeds and grasses serve as hosts for root knot nematodes. If the weeds are not controlled, the marigolds may be unable to suppress the nematode population.

Using marigolds in a manner other than that recommended can invite disaster. For example, planting a few marigolds here and there among tomatoes will encourage spider mites. The spider mite is one of the most difficult garden pests to control and can become nearly as serious a problem as the nematodes.

All nematode-infected plants should be removed from the garden as soon after production ceases as possible. Removing the root systems will eliminate many of the nematodes. To ensure that as much of the root system as possible is removed, use a shovel rather than just pulling the planting up by the stem.

Root knot resistant vegetable varieties are not plentiful. Fortunately, progress has been made in the development of root knot resistant tomatoes. The best resistance is found in the hybrid varieties that have been developed in recent years. Root knot resistant varieties are noted in seed catalogs by "N" following the variety name. Preceding the "N" are often "V" and "F." "VFN" stands for verticillium wilt resistance, fusarium wilt resistance, and nematode resistance. Hybrid tomato seed is more expensive than open-pollinated seed, but the benefits in disease resistance alone are enough to make it worth buying.

Diseases caused by root knot nematodes and other organisms can be prevented with non-chemical methods that either avoid the disease or suppress the disease agent. However, maximum control can only be assured by utilising as many of these disease preventive practices as possible.

REFERENCES

Matthews, R. E. Diagnosis of Plant Virus Diseases. Boca Raton, Florida: CRC Press, 1993.

Streets, Rubert B. The Diagnosis of Plant Diseases: A Field and Laboratory Manual Emphasizing the Most Practical Methods for Rapid Identification. Tucson: University of Arizona Press, 1972.

Vidhyasekaran, P. Concise Encyclopedia of Plant Pathology. New York: Food Products Press, 2004.

Westcott, Cynthia, revised by R. K. Horst. Westcott's Plant Disease Handbook. Boston : Kluwer Academic Publishers, 2001.

Bibliography

Agrios, George N., *Plant Pathology,* Fifth ed. Amsterdam: Elsevier, 2005.

Alfieri S.A Jr., El-Gholi N.E., "Phytophthora leaf blight of flowering dogwood, Cornus florida L.", *Plant Pathology Circular Gainesville,* 1993.

Bailey, L. H., *Manual of cultivated plants,* The Macmillan Company, Now York.

Bradbury, J. F., *Guide to Plant Pathogenic Bacteria,* Farnham Royal, England: CAB International, 1986.

Britton K.O, Redlin S.C., "Damping-off of flowering dogwood seedlings caused by Colletotrichum acutatum and Fusarium oxysporum", *Plant Disease,* 1995.

Browne, F. G. "Pests and Diseases of Forest Plantation Trees", *An Annotated List of the Principal Species Occurring in the British Commonwealth,* Oxford, England: Clarendon Press, 1968.

Canfield M.L, Baca S, Moore L.W., "Isolation of Pseudomonas syringae from 40 cultivars of diseased woody plants with tip dieback in Pacific Northwest nurseries", *Plant Disease,* 1986.

Forsberg, Junius L., *Diseases of Ornamental Plants,* University of Illinois. Urbana-Champaign, Illinois, 1975.

Fortune M and Dilbar A., "New diseases of mango (Mangifera indica), passion fruit (Passiflora edulis var flavicarpa) and pommecythere (Spondias cytherea) and their threat to sustainable agriculture", *Agriculture in a changing economic environment: Proceedings of the Seventh Annual Seminar on Agricultural Research,* San Fernando, Trinidad and Tobago, 9-10 November 1993.

George Agrios, *Plant Pathology,* New York: Academic Press, 1969.

George N. Agrios, *Plant Pathology,* fourth edition, Academic Press, New York, 1997.

Grimm, William C., *How To Recognize Shrubs,* Castle Books, NY, 1966.

Hibben C.R., "Anthracnose threatens the flowering dogwood", *Methods for diagnosing and controlling this new disease,* Arnoldia Boston, 1990.

Hottes, Alfred C., *Climbers and Ground Covers,* A. T. De La Mare Company, Inc., New York,1947.

Jeffries P and Jeger M J., "The biological control of postharvest diseases of fruit", *Biocontrol News and Information,* 1990.

Jerral D. Johnson and David N. Appel, *Major Oak Diseases and Their Control,* College Station: Texas Agricultural Extension Service, 1984.

Johnson, Warren T. and Lyon, Howard H., *Insects That Feed on Trees and Shrubs,* Cornell University Press. Ithaca, New York, 1976.

Kezer, A., and W. G. Sackett, *Beans in Colorado and their diseases,* 1918.

MacKenzie, David S., *Perennial Ground Covers,* Timber Press. Portland, Oregon, 1997.

Matthews, R. E. Diagnosis of Plant Virus Diseases. Boca Raton, Florida: CRC Press, 1993.

McRitchie J.J., "The felt fungus, Septobasidium", *Plant Pathology Circular Gainesville,* 1991.

Melle, P. J. Van, *Shrubs and Trees for the Small Place,* Doubleday and Company, Inc. Garden City, New York, 1955.

Nordbo, E., Christensen, S., Kristensen, K., Walter, M., "Patch spraying of weed in cereal crops", *Aspects of Applied Biology,* 1994.

Nordmeier, H. & Dunker, M., "Variable weed densities and soil properties in a weedmapping concept for patchy weed control", *In:Precision Agriculture '99,* Odense, Denmark, Seccond European Conference on Precision Agriculture, 1999.

Nyvall, Robert F., "Field Crop Diseases Handbook", Ames, IA: Iowa State University Press, 1999.

Purcell, A. H., "Patterns and processes of plant diseases and insect vectors", *Agriculture and Forestry Bulletin of the University of Alberta 5(3):19-22.*

Sherald J.L, Stidham T.M, Hadidian J.M, Hoeldtke J.E., "Progression of the dogwood anthracnose epidemic and the status of flowering dogwood in Catoctin Mountain Park", *Plant-Disease,* 1996.

Sinclair, Wayne A.; Lyon, Howard H.; Johnson, Warren T., *Diseases of Trees and Shrubs,* Cornell University Press. Ithaca, New York, 1996.

Streets, Rubert B. The Diagnosis of Plant Diseases: A Field and Laboratory Manual Emphasizing the Most Practical Methods for Rapid Identification. Tucson: University of Arizona Press, 1972.

Thompson, H. C., *Vegetable crops,* McGraw-Hill Book Company, Inc., New York. 1923.

Vidhyasekaran, P. Concise Encyclopedia of Plant Pathology. New York: Food Products Press, 2004.

Wendell Horne C. et al., *Texas Plant Disease Handbook,* College Station: Texas Agricultural Extension Service, 1983.

Westcott, Cynthia and Horst, R. Kenneth, *Westcott's Plant Disease Handbook,* Van Norstrand Reinhold Company, New York, 1978.

Whitwell A., "Dominica orchard crop management and research project: pest and disease management", *Final report: minor Tropical fruit crops,* Kent, United Kingdom: Natural Resources Institute, 1991.

Index